Debating the Faith: Religion and Letter Writing in Great Britain, 1550-1800

ARCHIVES INTERNATIONALES D'HISTOIRE DES IDÉES

INTERNATIONAL ARCHIVES OF THE HISTORY OF IDEAS

209

DEBATING THE FAITH: RELIGION AND LETTER WRITING IN GREAT BRITAIN, 1550-1800

Edited by

Anne Dunan-Page • Clotilde Prunier

For further volumes:
http://www.springer.com/series/5640

Anne Dunan-Page • Clotilde Prunier
Editors

Debating the Faith: Religion and Letter Writing in Great Britain, 1550-1800

Springer

Editors
Anne Dunan-Page
Aix-Marseille Université
LERMA (EA 853)
Institut Universitaire de France
Aix-en-Provence, France

Clotilde Prunier
CREA (EA 370)
Université Paris Ouest
Nanterre La Défense
Nanterre, France

ISSN 0066-6610
ISBN 978-94-007-5215-3 ISBN 978-94-007-5216-0 (eBook)
DOI 10.1007/978-94-007-5216-0
Springer Dordrecht Heidelberg New York London

Library of Congress Control Number: 2012952676

Printed on acid-free paper

Springer is part of Springer Science+Business Media (www.springer.com)

Contents

Notes on Contributors

Reid Barbour is Professor of English at the University of North Carolina, Chapel Hill. He has written on a wide range of subjects, including Elizabethan fiction, classical reception, Stuart religion, John Selden, Lucy Hutchinson, and Sir Thomas Browne.

Anne Dunan-Page is Professor of Early-Modern British Studies at Aix-Marseille University and a member of the Institut Universitaire de France. Her research interests include Restoration and early eighteenth-century dissent. She is currently working on the records of gathered Churches. Her latest publication is *The Cambridge Companion to Bunyan* (2010).

David Finnegan is currently teaching at Goldsmiths, University of London and at Rugby school. He has published several essays on Irish Catholics and Catholicism in the early-modern period and is currently completing a monograph based on his doctoral dissertation entitled *The Politics of Irish Catholicism, 1534–1653* (forthcoming). He has also co-edited two essay collections, *The Flight of the Earls* (2010) and *Varieties of Seventeenth- and Early Eighteenth-Century English Radicalism in Context* (2011).

Mark Greengrass is currently Senior Research Fellow at the School of History in the Freiburg Institute for Advanced Studies (FRIAS), Germany. He has worked for many years on the history of the Protestant Reformation, especially in France and England. From 2004 until 2009, he directed the British Academy John Foxe Project, based at the University of Sheffield, where he held a Personal Chair of Early Modern History. He is currently completing the Penguin History of Europe, vol. V (1517–1648) and working on a bigger project, focusing on epistolary culture in France, c. 1550–1643 entitled: 'Power and Communication in Renaissance France'.

Sarah Hutton is a graduate of New Hall, Cambridge and the Warburg Institute, London. She currently holds a chair at the University of Wales, Aberystwyth. Her main area of research is seventeenth-century intellectual history, with a special interest in the Cambridge Platonists. Her publications include, *Anne Conway. A Woman Philosopher* (2004). *Platonism at the Origins of Modernity* (edited with

Douglas Hedley, 2008), *Newton and Newtonianism* (edited with James E. Force, 2004), *Women, Science and Medicine* (edited with Lynette Hunter, 1996). She is Director of the series *International Archives of the History of Ideas*.

James Moore is Emeritus Professor of Political Science at Concordia University. Among his many positions, he has been President of the Conference for the Study of Political Thought and President of the Eighteenth-Century Scottish Studies Society. A specialist of eighteenth-century Scotland, he has been working for some years on an edition of the correspondence of Francis Hutcheson (1694–1746). He has edited, among others, *Natural Rights on the Threshold of the Scottish Enlightenment: The Writings of Gershom Carmichael* (2002, with Michael Silverthorne), *Francis Hutcheson on Logic, Metaphysics and Natural Sociability* (2006, with Michael Silverthorne) and *Francis Hutcheson on The Meditations of the Emperor Marcus Aurelius Antoninus* (2008, with Michael Silverthorne).

Claire Preston is Professor of Early-Modern Literature at the University of Birmingham. Recent books include *Thomas Browne and the Writing of Early-Modern Science* (winner of the British Academy's Rose Mary Crawshay Prize 2005) and, with Reid Barbour, *Sir Thomas Browne: The World Proposed* (2008). She is the General Editor of *The Complete Works of Sir Thomas Browne* (Oxford). She has held research awards from the British Academy, the Guggenheim Foundation, and Yale University. She is completing a book on early-modern literature and scientific investigation.

Clotilde Prunier is Professor in British history at the Université Paris Ouest Nanterre La Défense. She is the author of *Anti-Catholic Strategies in Eighteenth-Century Scotland* (2004), which focuses on the struggle between Presbyterians and Catholics in eighteenth-century Scotland, and in particular on how schools were used in the anti-Catholic crusade in the Highlands. She has published articles on eighteenth-century perceptions of the Highlands, on the role of education in eighteenth-century Scotland, and on Scottish Catholic correspondence. Her research has come to focus on the European correspondence of the clandestine Scottish Catholic community in the eighteenth century. She is currently working at a monograph on the continental networks—both human and epistolary—which enabled the Scottish Catholic Mission to weather the penal laws in the eighteenth century.

Gary Schneider is Associate Professor in the Department of English, University of Texas. He is the author of *The Culture of Epistolarity: Vernacular Letters and Letter Writing in Early Modern England, 1500–1700* (2005) and is currently working on a monograph entitled, 'Print Letters and Politics in Sixteenth- and Seventeenth-Century England.'

Alison Searle was a Research Associate on the the AHRC-funded *Complete Works of James Shirley* at Anglia Ruskin University. She was also Visiting Research Fellow at Dr Williams's Centre for Dissenting Studies and at the Centre for Baptist History and Heritage, Regent's Park College, University of Oxford. Her monograph entitled *The Eyes of Your Heart: Literary and Theological Trajectories of Imagining*

Biblically was published in 2008. She is currently preparing an edition of *The Sisters* (1642) by James Shirley for Oxford University Press. She is also working, with Dr Johanna Harris, on a complete edition of Richard Baxter's correspondence.

Daniel Szechi is a graduate of the University of Sheffield and St Antony's College, Oxford, and was appointed Professor of Early Modern History at the University of Manchester in 2006. He is a Fellow of the Royal Society of Edinburgh and the Royal Historical Society. His books include: *The Dangerous Trade. Spies, Spymasters and the Making of Europe* (2010); with Paul Kléber Monod and Murray G. H. Pittock, *Loyalty and Identity. Jacobites at Home and Abroad* (2009); *1715. The Great Jacobite Rebellion* (2006); *George Lockhart of Carnwath 1689–1727: A Study in Jacobitism* (2002); *The Jacobites. Britain and Europe, 1688–1788* (1994); with Prof. G. Holmes, *The Age of Oligarchy: Pre-Industrial Britain 1722–1783* (1993); and *Jacobitism and Tory Politics, 1710–1714* (1984). He has also published articles in *Past and Present*, *English Historical Review*, *Scottish Historical Review*, *Journal of British Studies*, *Historical Journal*, *Catholic Historical Review*, *Parliamentary History* and *Studies in Church History*, and a number of essays in collective works.

Ann Thomson is Professor of British History and director of the 'Groupe de recherches en histoire intellectuelle' at Université Paris 8. Recent publications include: *Bodies of Thought: Science, Religion, and the Soul in the Early Enlightenment* (2008, awarded the 2009 National Research Prize, SAES/AFEA) and *Cultural Transfers: France and Britain in the Long Eighteenth Century*, edited with Simon Burrows and Edmond Dziembowski (2010).

Chapter 1
Introduction

Gary Schneider

1.1 Correspondences

Before she was executed for treason on 12 February 1554, Lady Jane Grey, the nine days' queen, wrote a letter to John Feckenham, a Catholic priest with whom Grey had discussed matters of faith. Grey also wrote another letter to her sister, Lady Catherine Grey, an exhortation to Catherine to continue to practice the Protestant faith. These two letters were printed after Grey's death together with a debate with Feckenham and Grey's scaffold speech as *An Epistle of the Ladye Jane ... to a Learned Man of late Falne from the Truth of Gods Most Holy Word ... also Another Epistle whiche She Wrote to Her Sister* (1554).[1] Although Grey likely had a future audience beyond her correspondents in mind, the letter may not have been written expressly for the press (Daybell 2006a, 172).[2] Yet the letters in *An Epistle of the Ladye Jane* have the distinction of being among the first spiritual letters composed by a woman to see print in England, and saw an enduring publication history, including publication as part of a 1792 biography, *Life & Death of Lady Jane Grey*, by Thomas Gibbons.[3]

[1] Place of publication of the early modern texts to which I refer is London unless otherwise indicated.

[2] Other letters of Lady Jane Grey of similar devotional tenor, for instance, were not printed at the time (Daybell 2006a, 172).

[3] These letters were first published by Gibbons in his *Memoirs of Eminently Pious Women*, vol. 1 (1777), 25–37, 39–41. See Edith Snook, *Women, Reading, and the Cultural Politics of Early Modern England* (Aldershot: Ashgate, 2005), 53n77, for more on the publication history of this material.

G. Schneider (✉)
Department of English, University of Texas – Pan American, Beverly Hills Lane, 610, Edinburg, TX 78542, USA
e-mail: garysch@panam.edu

A. Dunan-Page and C. Prunier (eds.), *Debating the Faith: Religion and Letter Writing in Great Britain, 1550-1800*, International Archives of the History of Ideas 209,
DOI 10.1007/978-94-007-5216-0_1, © Springer Science+Business Media Dordrecht 2013

Five years after Gibbons' book was printed, *A Pastoral Letter from a Minister to His Parishioners, Being an Earnest Exhortation to Them to Take Care of Their Souls* came out in its twentieth edition. First printed in 1699, this pastoral letter of an anonymous minister had accrued by 1797, besides the exhortation, a devotional exercise, a catechism, and prayers. Considering a publication history that in 1797 had spanned almost a century, it was plainly an enormously popular epistle among the great number of this sort of moral-didactic letter printed during the eighteenth century. The book was also translated into Welsh and German, 3,000 copies were sent to the English army in Holland in 1701, while later editions went out to the new world (Jacob 1996, 110; Jablonski 1822, 481; 'An Account' 1817, 375; Monaghan 2007, 420n25). It was also recommended reading on doctrine and practice for members of the Society for Promoting Christian Knowledge, and in 1708 acquired a companion piece, *The Christian's Daily Devotion … Being a Continuation of the Pastoral Letter from a Minister to His Parishioners*.[4]

Three years before *A Pastoral Letter from a Minister to His Parishioners* appeared in its last edition of the century, Congregationalist Elizabeth Wilson wrote to her sister Rebekah Bateman in 1794 that 'A loving letter is all the comfort one has when absent from each other,' and in a letter of the prior year hopes that 'the Lord make us good soldiers of Jesus Christ that we may war a good warfare' (qtd. in Whyman 2010, 142, 144). Elizabeth Wilson's correspondence with her sister embodies in it the struggle she endured for her nonconformity. She used letters to express and strengthen her faith as well as to participate in an epistolary community of Congregationalists. The letters were not intended for future reading audiences, were never meant to be printed—indeed, were never intended to circulate beyond the sisters' private epistolary circle; rather they served as personal spiritual vehicles where 'A truly Christian letter might … become a means of grace' (Whyman 2010, 144).

Two and a half centuries earlier John Philpot, imprisoned during the reign of Queen Mary, begins a letter to John Careless, 'My dearly beloved brother Careless, I have received your loving letters, full of love and compassion.' Philpot requests in the letter that Careless 'Commend me to all our faithful brethren; and bid them with a good courage look for their redemption, and frame themselves to be hearty soldiers in Christ' (Bickersteth 1837, 171, 172). Using language markedly similar to that employed by Elizabeth Wilson, Philpot maneuvers his letter to exhort and encourage others to whom he does not have direct personal access. Yet unlike the letters of Elizabeth Wilson, those of John Philpot were printed as Protestant propaganda, appearing in *Certain Most Godly, Fruitful, and Comfortable Letters of Such True Saintes and Holy Martyrs of God* (1564)—also known as *The Letters of the Martyrs*—a collection deliberately foregrounding in print epistolary communication within a conspicuous epistolary community. Philpot's letter was reprinted in the 1583 edition of John Foxe's *Actes and Monuments* where it continued to serve as religious propaganda.[5]

---

[4] See *An Account of the Society for Promoting Christian Knowledge* (1775), 70, for the recommendation.

[5] See John Foxe, *Actes and Monuments*, vol. 2 (1583), 1833.

I have selected these letters expressly to outline the contours of this collection, which spans the years 1550–1800, yet I do not wish to imply that during this period no systemic shifts in epistolarity occurred—without doubt, crucial changes in the postal system, the ascendancy of print culture over manuscript culture, and marked increases in literacy (augmenting the number of those who could read and write letters) characterize these two and a half centuries. Rather, I wish to accentuate continuity and persistence within a broad early modern letter-writing culture that preceded the Industrial Revolution when novel modes of immediate communication over distances such as the telegraph, radio, and telephone transformed the way the letter was employed and epistolarity was imagined—continuities I think even more marked in letters of a religious, spiritual, moral-didactic, and controversial nature of Catholics and Protestants alike. Such continuities include the sustained use of letters purely as vehicles of communication among individuals within a specific religious epistolary community, as evident in the letters of Elizabeth Wilson and John Philpot; it includes the ongoing composition of handwritten religious letters by women in manuscript—as demonstrated by Elizabeth Wilson—and potentially for print—as in the case of Jane Grey; the persistence of forms of religious rhetoric, revealed in the letters of Philpot and Wilson; the continuing dynamic relationship between manuscript and print cultures, evidenced in that *A Pastoral Letter* may have been written expressly for print, Philpot's and Grey's letters for the possibility of print, while the spiritual letters that Wilson wrote were never intended for publication; and the fundamental consistency among all these letters to manifest an exhortatory character intended to energize faith. Moreover, the use of letters exchanged in handwritten texts or printed for mass circulation to hash out religious controversy and to level animadversion continued, while the use of letters as components of spiritual and ecclesiastical biography developed fully during this period.

1.2 Background

In one of the very few studies to evaluate the influence of the Pauline letters on Renaissance epistolarity, C.A. Patrides states that 'formal letter-writing during the Renaissance was affected after some fashion by the self-evident importance of the epistle in the New Testament' (Patrides 1981, 357). His study accentuates the influence of the Pauline epistle on early modern letter-writing culture over the classical tradition, represented principally by the letters of Cicero, Seneca, and Pliny the Younger—who have been discussed far more often in the context of letter writing during this period than the New Testament epistolists. John Donne, for instance, writing to Henry Goodyear around 1604, lists who he considers the important letter writers of the past—Seneca, Pliny the Younger, Cicero, Phalaris, Brutus, and St. Paul—all but one classical writers (Donne 1651, 105–6). Erasmus himself, *the* authority on letter writing of the European Renaissance, favored classical models of letter writing over others, at least in terms of epistolography: Cicero and Pliny the Younger are the classical letter writers to whom Erasmus most frequently refers, while the letters of Church Fathers such as Jerome, Cyprian, and Augustine are mentioned far less often (Gerlo 1971, 111). Yet Church Fathers such as Jerome proved influential on epistolary writing in the Renaissance, while the letters of Augustine were prepared with an eye to future readers as patristic letter writers undoubtedly influenced future generations of letter writers; and the epistles of Paul were, of course, analyzed frequently, often at great length, by many early modern writers (Constable 1976, 28, 30; Patrides 1981, 361–5).

Perhaps it is no surprise, therefore, that the vast majority of modern scholarship on early modern letters and letter writing has likewise emphasized the classical tradition over the Christian tradition. Of course, there is no doubt that the classical legacy of letter writing, as Patrides points out, deserves considerable analysis. The work of Judith Rice Henderson is a good example of some of the fine scholarship on Renaissance epistolarity done with attention to the impact of classical letter writing.[6] Yet research on letters and letter writing detailing the impact of received tradition has betrayed an imbalance by concentrating on the authority and influence of classical letter writers.

During the late 1990s and into the new century, however, literary and historical scholarship on the early modern period has 'turned' to religion, as Ken Jackson and Arthur Marotti have termed it. Whereas race, class, and gender (and associated imperialist, materialist, and feminist approaches) have generally dominated literary scholarship since the advent of New Historicism, religion—though often implicit in such approaches—has generally either been forced aside or else subsumed within political, social, or economic paradigms; furthermore, historians have tended to elide religion and politics, or else read religion as a coded form of social, economic, and political phenomena (Jackson and Marotti 2004, 167–8).

The same can be said specifically of scholarship on letters and letter writing of the early modern period. Because spiritual, moral-didactic, and homiletic letters have a long pedigree stemming from Paul's epistles, religion has always been intrinsic to discussions of letters from the early modern period, but is often scattered and diffused within other categories. James Daybell's systematic bibliographies of scholarship on sixteenth- and seventeenth-century letters, for instance, do not contain a specific category for 'religion and letter writing' or 'religious letters'; rather they are classed within groupings of 'specific individual letter writers,' 'women's letters,' and 'genre studies,' which is a plain indication that the category of 'letters and religion' is difficult to define precisely because it is so pervasive (Daybell 2005, 2006b). Similarly, Daybell writes that 'Interest in letters and letter-writing has led to a range of truly interdisciplinary inquiries, including literary, lexical, historical, social, cultural,

[6] Henderson, 'Defining the Genre of the Letter: Juan Luis Vives' *De Conscribendis Epistolis*,' *Renaissance and Reformation* n.s. 7 (May 1983): 89–105; 'Erasmus on the Art of Letter-Writing' in James J. Murphy (ed.), *Renaissance Eloquence: Studies in the Theory and Practice of Renaissance Rhetoric* (Berkeley, Los Angeles, London: Univ. of California Press, 1983), 331–55; and 'On Reading the Rhetoric of the Renaissance Letter' in Heinrich F. Plett (ed.), *Renaissance-Rhetorik/Renaissance Rhetoric* (Berlin, New York: de Gruyter, 1993), 143–61, serve as representative examples.

intellectual, paleographical, manuscript, and gender-based approaches. Letters have been read as literary and cultural texts, and also as social documents' (Daybell 2005, 349–50). That no specific mention of 'religion' occurs is, again, a result of broader movements in modern scholarship on which Jackson and Marotti have commented—and which are presently being addressed. In a separate study Daybell, for instance, has also recognized that 'The history of the letter as a religious site can be traced back to biblical example and apostolic letters' (Daybell 2009, 522).

1.3 Letters and Religion, 1550–1800

The task of systematizing and organizing 'religion and letter writing' in the 1550–1800 period is no simpler now than it has ever been. The range of documents is enormous and the religious subject matter exceedingly broad. Letters, both those exchanged in handwritten texts and those printed during the early modern period, served a countless number of religious and spiritual functions during this time, and, while some of these continued uses common in Europe before the early modern era, others proved to be unique innovations of epistolarity.

Perhaps most frequently, sermons were often transmitted by letters, letters that were exhortatory, consolatory, and advisory in nature. To take but a handful of examples of the many representative illustrations, Lady Rachel Russell's epistolary exchange with Dr. John Fitzwilliam between the years 1680 and 1696 constitutes a correspondence whose primary purpose was for Fitzwilliam to exhort, console, and advise his correspondent Russell.[7] The extensive spiritual letters of non-conformist preacher William Huntington (1745–1813) to family, friends, and parishioners are also homiletic and exhortatory in nature.[8] Sermons as letters shade into advisory letters when such letters counsel on spiritual matters. Henry More sent such letters to correspondent Anne Conway, while women such as Anne Bacon, Katherine Paston, and Brilliana Harley wrote letters of spiritual advice to their sons.[9]

After the advent of print in England, sermons as letters began to be published to reach mass audiences. Letters as sermons were often printed posthumously, as in *The Christian Letters of Mr. Paul Bayne* (1620), but sometimes non-posthumously: the

[7] See *Letters of Lady Rachel Russell* (1773).

[8] See Huntington's correspondence in the British Library, Additional Manuscript 46886.

[9] See Sarah Hutton (rev. ed.), *The Conway Letters: Correspondence of Anne, Viscountess Conway, Henry More, and Their Friends* (Oxford: Clarendon, 1992); Ruth Hughey (ed.), *The Correspondence of Lady Katherine Paston 1603–1627* (Norfolk: Norfolk Record Society, 1941); Thomas Taylor Lewis (ed.), *Letters of the Lady Brilliana Harley* (London: Camden Society, 1854). The scholarship includes Lynne Magnusson, 'Widowhood and Linguistic Capital: The Rhetoric and Reception of Anne Bacon's Epistolary Advice,' *English Literary Renaissance* 31.1 (Winter 2001): 3–33; and Raymond A. Anselment, 'Katherine Paston and Brilliana Harley: Maternal Letters and the Genre of Mother's Advice,' *Studies in Philology* 101.4 (Autumn 2004): 431–53.

letters in the first vernacular collection published in England by a living native Englishman, for instance, Joseph Hall's *Epistles* (1608, 1611), take the sermon-as-letter form. At 214 folio pages, *An Epistle of Comfort to the Reverend Priests* (1587) by Robert Southwell broadens the sermon-as-letter form into a martyrological discourse, while other sermons as letters, such as Robert Cottesford's *Two Very Godly and Comfortable Letters ... One to a Godly and Zealous Lady wherin the Annabaptists Errour is Confuted ... The Other an Answer to a Godly Merchants Letter* (1589) challenge the boundary between consolatory homily and religious polemic. *The Bishop of Hereford's Pastoral Letter to the Inhabitants of His Diocese* (Hereford, 1798) demonstrates that this frequent usage continued to the end of the eighteenth century.

Clerics also employed letters in associated ways. One usage consists of the letter to engage debate, controversy, and animadversion since the letter—based as it is on the paradigms of obligation, reciprocity, and exchange—ideally suited these practices. A correspondence between William Bedell and James Wadsworth, who had befriended one another in Cambridge and later both held livings in Suffolk, took place between 1615 and 1620 after Wadsworth had converted to Catholicism and left England for Spain.[10] The correspondence was not intended for print, but in 1624 it was published as an ongoing religious debate entitled *Copies of Certaine Letters which Have Passed betweene Spaine and England in Matter of Religion ... betweene Master James Wadesworth ... and W. Bedell.* Other collections inhabit the indistinct space between 'intended for print/not intended for print,' such as the doctrinal debate in epistolary form between John Jewel and Henry Cole, *The True Copies of the Letters betwene ... John Bisshop of Sarum and D. Cole* (1560). Other letters prepared for the press to engage debate, controversy, and animadversion during the eighteenth century include *A True Copy of a Letter lately Written by Mr. Dowley to Dr. Wells ... together with the Doctor's Answer* (Oxford, 1706) and *Letters which Passed between the Right Reverend Robert Lord Bishop of Corke ... and Mr. William Penn, Concerning Baptism* (1756).

Ecclesiastics also continued in the early modern era to use the letter in Church organization, in determining articles of faith, in discussing biblical exegesis, and in gathering news and intelligence. The letters exchanged by reformers John Jewel, Peter Martyr, Henry Bullinger, John Parkhurst, and Edmund Grindal demonstrate for instance an epistolary community concerned with all of these activities.[11] Father Anthony Rivers wrote letters of religious and other news to correspondent Robert Persons, some of which correspondence was intercepted; Richard Verstegan was the center of a considerable Catholic correspondence network, Henry Garnet and Robert Southwell numbering among his correspondents in England, while a variety of informants and recipients on the continent—including Persons among the latter—sent

[10] *Oxford Dictionary of National Biography*, sub nom. Bedell, William; Wadsworth, James.

[11] See Hastings Robinson (ed. and trans.), *The Zurich Letters Comprising the Correspondence of Several English Bishops and Others*, 2 series (Cambridge: Cambridge Univ. Press, 1842–1845) and *Original Letters Relative to the English Reformation*, 2 vols (Cambridge: Cambridge Univ. Press, 1846–1847).

information to and received information from Verstegan (Arblaster 2004, 68, 70).[12] It has also been recognized how astutely the Quakers used letters, both in manuscript and print, to communicate, organize, and petition.[13] In the eighteenth century, handwritten letters helped Methodists carve out a distinct evangelical space for themselves; John Telford has called John Wesley's letters 'the marching orders of the Evangelical Revival' (Brant 2006, 312; Telford qtd. in Brant 2006, 282).

Letters from prison have always been a common sub-genre of epistolary writing, no more so than in religious contexts—Paul's epistles to the Ephesians, Philippians, Colossians, and Philemon mark the Christian archetype. That Paul was imprisoned for his ministry was a fact, for example, not lost on Nicholas Ridley writing to the imprisoned John Bradford:

> since they [those who have imprisoned you] have changed their purpose, and prolonged [i.e., deferred] your death, I understand it is no other thing than that once happened to Peter and Paul. The which, although they were of the first which were cast in prison ... yet God would not have them put to death with the first, because he had more service to be done by their ministry (Bickersteth 1837, 47).[14]

The Marian martyrs indeed wrote a number of letters from prison, as did Puritan separatist John Penry later in the century, while Mary Ward, founder of the Institute of the Blessed Virgin Mary, wrote letters in lemon juice on wrapping paper while imprisoned (Littlehales 2001, 206)—to give but a handful of instances. This type of letter swiftly migrated into print. Thomas More's prison letters of 1534–1535 were printed in 1557 in his *Englysh Workes*, but this type of gathering culminated relatively early (in 1564) in the massive collection, *Certain Most Godly, Fruitful, and Comfortable Letters*, which effectively comprises over 200 letters to and from imprisoned Protestants, which, as a collection of prison letters of multiple hands, was never to be duplicated in size during this historical period. The mid-seventeenth century saw Presbyterian Christopher Love's prison letters printed in *Love's Name Lives* (1651) and Leveller John Lilburne's in *A Coppy of a Letter Written by John Lilburne, Close Prisoner in the Wards* (London?, 1640), while *A Letter Written by Mr. John Dickson, (Late Minister of the Gospel at Rutherglen) from the Bass Prison* (Edinburgh?, 1719) exemplifies the persistence of this sort of publication into the following century.

Another textual practice, one that developed fully as a function of print culture, consists of the letter as 'relic,' where one's literary remains were printed to serve as textual (rather than as corporeal) memorials of that individual, either in the context

[12] See Henry Foley (ed.), *Records of the English Province of the Society of Jesus*, vol. 1 (London: Burns and Oates, 1875), 4–66, for Rivers' letters.

[13] See, for instance, Kate Peters, 'Patterns of Quaker Authorship, 1652–1656' in Thomas N. Corns and David Loewenstein (eds.), *The Emergence of Quaker Writing: Dissenting Literature in Seventeenth-Century England* (London: Frank Cass, 1995), 6–24; and Matthew Horn, 'Texted Authority: How Letters Helped Unify the Quakers in the Long Seventeenth Century,' *Seventeenth Century* 23.2 (Autumn 2008): 294–318.

[14] See also Bickersteth 1837, 378, 380; and Sarah Covington, *The Trail of Martyrdom: Persecution and Resistance in Sixteenth-Century England* (Notre Dame, IN: University of Notre Dame Press, 2003), 96–102.

of martyrdom or otherwise. This phenomenon was particularly true of Churchmen, where letters perform this function in, for instance, *The Genuine Remains of That Learned Prelate Dr. Thomas Barlow* (1693), yet *Reliquiae Sacrae Carolinae, or the Works of That Great Monarch and Glorious Martyr King Charls the I* (1650) includes a large amount of epistolary material and unmistakably celebrates the king's martyrdom and sanctification. On the other hand, *The Speeches, Discourses, and Prayers of Col. John Barkstead, Col. John Okey, and Mr. Miles Corbet* (1662)—which also includes letters—served to frame the printed relics of executed republicans as quasi-religious documents by tying these men to those that suffered for the primitive Church: just as 'the primitive Christians long[ed] for Martyrdom, seeking and pursuing after it: publicly and boldly to the very Teeth of the Tyrants and Persecutors proclaiming their Christianity,' so these men were martyred for supporting the Christian Commonwealth.[15]

Of course, the line between memorial and propaganda—as with the line between sermon and polemic—was not a clear one, since letters were utilized in a number of different ways in print to reach a wide reading public on matters of faith and religion. William Cecil, Lord Burghley, for instance, engaged in a propagandistic 'epistolary fiction' when taking the role of an English Catholic loyal to Queen Elizabeth in his *The Copie of a Letter sent out of England to Don Bernardin Mendoza* (1588). Richard Verstegan in his equally fictive print letter, *The Copy of a Letter lately Written by a Spanishe Gentleman* (Antwerp, 1589), responded to this pamphlet by taking the role of a Spanish gentleman newly freed from captivity after the failure of the Armada, who, in writing to his friend in England, has cause to dispute the content of *The Copie of a Letter Sent out of England to Don Bernardin Mendoza*. This sort of religious epistolary fiction persisted into the next centuries, for instance, in *A Letter from Father La Chaise, Confessor to the French King, to Father Peters* (1688), where François de la Chaise is ventriloquized, while similar forged letters as if by clergymen, such as *Copy of a Letter Addressed to the Father Rector at Brussels, Found among Some Jesuits Taken at London* (1643), masquerade as real 'discovered' letters. Yet other such letters are satirical in nature, such as the ventriloquized *The Popes Letter to Maddam Cellier* (1680) and Elizabeth Cellier's pretended response, *Maddam Celliers Answer to the Popes Letter* (1680).

Letters written by women alert us to the cultural meanings of gender and its impact on religious letter writing of all types. Throughout the period, moral-didactic subjects were generally ones on which it was acceptable for women to write and—eventually—in which to publish. Handwritten letters by women range from those seeking support in correspondence with ministers such as John Knox, Edward Dering, Thomas Cartwright, and Thomas Wilcox, to those of a confessional nature written by individuals such as Margaret Clifford to her chaplain,

[15] *Speeches, Discourses, and Prayers*, To the Reader, unsigned.

while other women such as Margaret Neville and Elizabeth Beaumont employed letters pragmatically when responding to charges of recusancy, often using gendered rhetoric in doing so (Daybell 2009, 524, 526, 529–32). Published religious letters by women were relatively few, and those chiefly printed posthumously, as in *An Epistle of the Ladye Jane*, but by the mid-seventeenth century living women letter writers published somewhat more often. Quaker women, among them Mary Howgill, Margaret Fell Fox, Anne Docwra, and Anne Gilman, engaged the press to print letters on behalf of the Friends, some of which were addressed to political leaders.[16] Later in the century Mary Astell in *Letters Concerning the Love of God between the Author of the Proposal to the Ladies and Mr. John Norris* (1695) demonstrated the growing phenomenon of women printing moral-didactic letters, a development that continued in the eighteenth century with writers such as Hester Chapone, who published the exceedingly popular *Letters on the Improvement of the Mind, Addressed to a Young Lady* (1773). Yet other moral-didactic letters in print—but ones never intended for print when composed—were published during the period, such as those included in *The Holy Life of Mrs. Elizabeth Walker, Late Wife of A[nthony] W[alker] D. D. Rector of Fyfield in Essex* (1690), in which six letters of a consolatory and exhortatory nature were included, set off from the biography proper.

Spiritual biography, biographically oriented martyrdom accounts, and biographies of ecclesiastics exploited letters during this period, as well. John Mush's 'Life of Margaret Clitherowe' contains a letter written by the Lord Mayor of York on behalf of Margaret's husband to the Earl of Derby seeking their daughter Anne's release from prison, among other documents. Jesuit John Gerard composed his 'life' at the request of his superiors, an autobiographical account in which Gerard quotes portions of letters in the narrative.[17] Other biographies printed during the early modern era incorporated letters, either threaded through the biography such as in William Newton's *The Life of the Right Reverend Dr. White Kennett, Late Lord Bishop of Peterborough* (1730) and William Gilpin's *The Life of Thomas Cranmer, Archbishop of Canterbury* (1784) or else set apart from the biography proper, as in Edward Smyth's *The Extraordinary Life and Christian Experience of Margaret Davidson* (Dublin, 1782) and Richard Parr's *The Life of the Most Reverend Father in God, James Usher, Late Lord Arch-Bishop of Armagh* (1686), to which Parr appends over 300 of Ussher's letters. The inclusion of letters in all such biographies points to the increasing use of letters as documentary evidence in religious contexts.

---

[16] Mary Howgill's *A Remarkable Letter of Mary Howgill to Oliver Cromwel* (1657) and Anne Gilman's *An Epistle to Friends ... also a Letter to Charles, King of England* (1662) serve as two instances.

[17] See John Morris (ed.), *The Troubles of Our Catholic Forefathers Related by Themselves*, vol. 3 (London: Burns and Oates, 1877), 354; and John Morris (ed.), *The Life of Father John Gerard, of the Society of Jesus* (London: Burns and Oates, 1881), 284–5 and 378, for instance.

Many forms of epistolary rhetoric likewise persisted throughout the centuries. The language of epistolary propriety, reliability, authenticity, and that concerning postal processes such as delivery and reception are perhaps the most salient, but religious rhetoric is equally enduring in letters, as the shared usage of the language of spiritual warfare in the letters of John Philpot and Elizabeth Wilson demonstrate, where Philpot uses his letter to frame the fight for reformed religion against Catholicism, while Wilson exploits the same rhetoric to image Congregationalism's struggle against Anglicanism. The language is likewise employed by Robert Southwell in his *Epistle of Comfort*: 'Your lyfe is a warfare your weapons patience, your Captayne Christe, your standerd the Crosse. Now is the larum sounded, and the warre proclaymed' (Southwell 1587, 134v). Other forms of religious rhetoric found in letters include the rhetoric of suffering and martyrdom, also common to Catholics and various Protestant denominations alike. Secular priest William Hart uses the rhetoric when writing to his mother in 1584: 'I dy … onely for my faith … for my blessed Saviour Jesus Christ…. How glad then may he bee to see mee a martyr, a Saint, a most glorious and bright starre in heaven' (qtd. in Marotti 2000, 175 [third set of ellipses in original]), while in his farewell letter composed just before his execution, Nicholas Ridley writes that Christ 'hath vouchsafed to call me … unto this high dignity of his true prophets, of his faithful apostles, and of his holy, elect, and chosen martyrs; that is, to die, and to spend this temporal life in the defence and maintenance of his eternal and everlasting truth' (Bickersteth 1837, 61). The language of communion through Christ, sometimes in gustatory metaphor, is another sort of spiritual rhetoric. John Bradford begins a letter, 'I heartily commend me unto you in our common Christ: whom I so call, not that I would make him as common things be, that is, nothing set by; but because by him we are brought into a communion' (Bickersteth 1837, 294). Edmund Waller writing to fellow Quaker Henry Goulding echoes this language roughly a century and a half later: 'Thy letter of the 23rd inst. was extreame wellcome to me. A sober felowship and communion in Christ Jesus is food and refreshment to the soules of the Righteous' (Locker-Lampson 1910, 69). John Wesley writes similarly to Jane Bisson in 1787, beginning the letter, 'I have a great union of spirit with you,' and continues, 'What you speak of your communion with Him comforts and warms my heart. I love to read or to hear any part of your experience' (Wesley 1960). In these instances, the rhetorical drive of the letters is to unite writer, recipient, and Christ in communication/communion, a drive plain in Methodist letters like Wesley's where 'Letters played out etymological connections between communion, communication, and community' (Brant 2006, 313). Now, without doubt these various rhetorical strategies appear in religious discourses besides letters; yet their manifestation in correspondence—exemplified especially in the last set of examples—demonstrates the extent to which this language had considerable value in stimulating devotion and strengthening communion among the faithful outside of formal or ceremonial discourses (such as sermons, catechisms, and prayer books) and organized for a such as Church services. At the same time, the inscription of this rhetoric in letters suggests an internalized (and personalized) consciousness of tropes of spirituality found precisely in other sorts of religious texts.

1.4 The Current State of Scholarship on Religion and Letter Writing

Modern scholarship that deals in some way with early modern British letters and religion over roughly the last 40 years may be generally divided into five groups: (1) studies of women's letters and religion, (2) general studies of epistolary writing that in part investigate the relationship between religion and letters, (3) literary and historical studies of religious topics in which letters are examined or figure significantly in the analysis, (4) studies of individual Churchmen and their letters, and (5) religion in the personal letters or in the epistolary fictions of literary writers.[18]

The largest category of the secondary literature consists of research on women's letters and religion, and is directly associated with the rise of feminist scholarship. There are no book-length studies that deal solely with British women, letters, and religion, but numerous article- and chapter-length studies as well as monographs deal in part with women, letters, and religious subject matter. In general, this scholarship has revealed to what remarkable extent women of all religious denominations participated in the culture of epistolarity, the dexterous and multi-faceted uses to which letters were put, and the various contexts in which women wrote—from private, spiritual ones to public, controversial ones, the research spanning from letter writers such as Elizabeth Talbot, Countess of Shrewsbury, writing just after the mid-sixteenth century, to Eliza Gould, writing at the end of the eighteenth.

Monographs that explore early modern epistolary writing generally that in part investigate the relationship between letters and religion are fewer. Thomas O. Beebee's *Epistolary Fiction in Europe, 1500–1850*, Clare Brant's *Eighteenth-Century Letters and British Culture*, James Daybell's *Women Letter Writers in Tudor England*, Gary Schneider's *Culture of Epistolarity: Vernacular Letters and Letter Writing in Early Modern England, 1500–1700* and Susan E. Whyman's *The Pen and the People: English Letter Writers, 1660–1800* are among them.[19] Each of these monographs speaks to religion and letter writing in some capacity, but even though Brant devotes an entire chapter to the topic of letter writing as a Christian, religion and letter writing are not the principal focus of any of these.

Literary and historical analyses of religious topics in which letters are examined or figure significantly in the analysis are exemplified by several monographs, among them Brad S. Gregory's *Salvation at Stake: Christian Martyrdom in Early Modern*

---

[18] I am indebted to Daybell's comprehensive bibliographies of sixteenth- and seventeenth-century letters and letter writing as these serve as the basis for the brief review of the 'religion and letter writing' scholarship I undertake here; I refer the reader to these bibliographies rather than list the numerous individual articles and book chapters here.

[19] Beebee, *Epistolary Fiction in Europe, 1500–1850* (Cambridge: Cambridge University Press, 1999); Schneider, *The Culture of Epistolarity: Vernacular Letters and Letter Writing in Early Modern England, 1500–1700* (Newark: University of Delaware Press, 2005). The others are included in the reference list.

Europe, which details the scribal communities of Protestants, Anabaptists, and Roman Catholics; John R. Knott's *Discourses of Martyrdom in English Literature, 1563–1694*, which tackles *The Letters of the Martyrs* and George Fox's *Epistles*; and Scott R. Pilarz's *Robert Southwell and the Mission of Literature, 1561–1595*, in which Southwell's *An Epistle of a Religious Priest unto His Father* (London?, 1597?) and his *Epistle of Comfort* are examined.[20] Additionally, several article- and chapter-length studies examine letters in the wider context of religious issues. Whereas letters are often used as standard documentary evidence in literary and historical research, in this scholarship letters themselves are given detailed analysis.

Studies of individual Churchmen and their letters are fewer in number and are chapter and article length. Edward Dering, John Sheterdon, and John Wesley are among the religious figures whose letters have been investigated in the scholarship, although a number of studies have focused on John Knox, in particular his correspondence with women. The fifth grouping, investigations of religion in the personal letters or in the epistolary fictions of literary writers, by contrast, constitutes a somewhat larger portion of scholarship on religion and epistolary studies. A number of these studies have focused on religion in and the spiritual features of John Donne's and Thomas More's letters, while others have examined religious dimensions in the epistolary work of Thomas Browne, Aphra Behn, Mary Wortley Montagu, and Hester Chapone.

1.5 Ongoing Correspondences: The Present Collection

The purpose of this collection, then, at its simplest, is to extend prior research as well as to remedy deficiencies in the current scholarship on religion and early modern letter writing by bringing together an array of essays from both historians and literary scholars. The research that constitutes this collection indeed reflects the range of scholarship on letters and religion that has begun to be written since the 'turn' to religion. This collection of scholarship on early modern letter writing, whose purpose is to focus exclusively on religion, is long overdue. Individually, each chapter advances prior research on early modern letters and religion, and shares a set of common concerns with other essays in the collection.

In Chap. 2, 'Scribal Networks and Sustainers in Protestant Martyrology,' Mark Greengrass addresses several of the issues with which modern scholarship has been engaged regarding religion and letter writing, concentrating on reformers' prison letters in print, namely the attention given to outlining, and sometimes purposefully manipulating, the presentation of the epistolary communities of the reformers in

[20] Gregory, *Salvation at Stake: Christian Martyrdom in Early Modern Europe* (Cambridge: Harvard University Press, 1999); Knott, *Discourses of Martyrdom in English Literature, 1563–1694* (Cambridge: Cambridge University Press, 1993); Pilarz, *Robert Southwell and the Mission of Literature, 1561–1595: Writing Reconciliation* (Aldershot: Ashgate, 2004).

these printed collections to enhance their portrayal of faith-based communication as community. Analysis of epistolary community also characterizes the following chapter, Reid Barbour's 'Thomas Browne, the Quakers, and a Letter from a Judicious Friend,' although Barbour turns from a group of letters to a single letter from Quaker Samuel Duncon to Anglican (and fellow Norwich resident) Thomas Browne, in which Duncon invites Browne into the circle of Friends after reading Browne's *Religio Medici*—which itself is framed as a response to a letter from a friend. As such, the letters, at least notionally, take part in a sort of 'debate' environment where orthodox and heterodox theological issues are hashed out—a phenomenon also investigated, as we shall see, in Ann Thomson's contribution. Richard Baxter served as the core of yet another sort of epistolary community, one in which Baxter used his letters to strengthen his pastoral authority. This is the topic of Chap. 4, Alison Searle's 'Writing Authority in the Interregnum: the Pastoral Letters of Richard Baxter,' in which Baxter's letters to four of his correspondents serve as case studies that demonstrate how Baxter used his letters as vehicles of pastoral power during the fragmentation of religious authority that characterized the Interregnum. Baxter's method in doing so is both rhetorical—in the distinctive exhortatory and consolatory language he employs in his letters—as well as structural—in the exploitation of both print and manuscript to develop and consolidate his spiritual authority.

Rather than outlining an epistolary community, Anne Dunan-Page instead delineates how extensively letters served in documentary, legal capacities in the Baptist congregation at Cripplegate in her chapter, 'Letters and Records of the Dissenting Congregations: David Crosley, Cripplegate and Baptist Church Life'—and in doing so highlights the vital function letters continued to serve in Church administration. The Cripplegate congregation is exceptional in how it preserved letters as part of a practice of recording congregation business. In David Finnegan's contribution "'For the Greater Glory": Irish Jesuit Letters and the Irish Counter-Reformation, 1598–1626,' letters also document, in a sense, the success of the Jesuit mission in Ireland; however, these Jesuits overstated in their letters for propaganda's sake the success of the Jesuit mission. In this regard, Finnegan's work and Mark Greengrass' essay serve as complements to one another, as Greengrass examines in part the production of Protestant print propaganda. Finnegan also confronts the uncritical acceptance of these letters by historians, who have taken the Jesuit letter writers' exaggerations for facts—a salutary observation about letters as documentary evidence in religious contexts that reminds one of how recurrently letters acted as documentation in these environments—exemplified as well, as I have noted, in Anne Dunan-Page's work.

In assessing letters from abroad, Daniel Szechi, in Chap. 7, 'Negotiating Catholic Kingship for a Protestant People: 'Private' Letters, Royal Declarations and the Achievement of Religious Detente in the Jacobite Underground, 1702–1718,' concentrates on the purpose formal letters served the exiled James to represent the spirit and character of his Catholicism to his Jacobite adherents—indeed, to a Protestant people—back in Great Britain, should he be restored to the throne. The fundamental *raison d'être* of letters—to communicate over distances—is exploited by James and his advisors abroad to construct carefully worded documents, which, at their core, served as both literal and symbolic contact between an 'absent monarch' and his people.

This essential purpose of letters is also developed by Clotilde Prunier in her contribution, "'Every time I receive a Letter from you it gives me new vigour": the Correspondence of Scalan Masters, 1762–1783,' since epistolary intercourse was the only mode of regular communication the masters of the Scalan seminary in Scotland had with their superiors and with former students on the continent. The value of the correspondence was twofold here: some letters were devoted to the organization of the mission, while others served the key religious function of spiritual direction and reinvigoration, a purpose, as Alison Searle indicates, that Richard Baxter's letters also served.

In her '"Utopian Intelligences": scientific correspondence and Christian virtuosos,' Claire Preston examines an epistolary community consisting of various members of the Royal Society who exchanged letters on scientific phenomena—as was customary in the Society. Preston's aim, however, is not to address the fact of exchange, but rather to pay attention to a specialized epistolary rhetoric in which these individuals positioned scientific inquiry; and just as persistent forms of religious rhetoric were shared, so some members of the Royal Society sought an appropriate epistolary language of science ideally to be shared among those engaged in scientific endeavor. Sarah Hutton, like Preston, also examines the republic of letters, but shifts focus to a specific woman participating in the republic of letters. This is the topic of Chap. 10, 'Debating the Faith: Damaris Masham (1658–1708) and Religious Controversy,' in which Hutton details Masham's engagement with religious controversy by way of epistolary debate, one of the principal uses of letters in matters of spirituality during the period; Masham, then, exemplifies the emerging late seventeenth-/early eighteenth-century female letter writer tackling controversy in both print and manuscript.

While Chap. 11, James Moore's 'Evangelical Calvinists versus the Hutcheson Circle: Debating the Faith in Scotland, 1738–1739,' returns to a specific debate context—like that of Damaris Masham and John Locke—Moore assesses printed letters intended to criticize Francis Hutcheson, professor of moral philosophy at the University of Glasgow, where letters were expressly employed to dispute matters of religious conviction; but whereas Masham and Locke's correspondence was private, the letters surrounding Hutcheson were purposefully published in order to make the issues at stake public; as such, the letters are exemplary instances of the frequent usage of the print letter as locus for public debate. By contrast, the situation of privacy was crucial for Jean Bouhier, as Ann Thomson asserts in the final chapter, 'Questioning Church Doctrine in Private Correspondence in the Eighteenth Century: Jean Bouhier's Doubts Concerning the Soul,' where the circulation of 'clandestine literature' depended upon secrecy and confidentiality. It was precisely in his private correspondence that Bouhier expressed unorthodox (even heretical) theological views that could never bear the publicity of print—of course a posture diametrically opposed to those who sought print precisely to publicize their letters for controversial or propagandistic objectives, such as those letters examined by James Moore.

The research gathered here demonstrates a comprehensive perception of religion and letter writing in the early modern period. The correspondences among these chapters clarify the patterns that scholarship on religion and letters is forming; at the same time, this research is indicative of the type of research that has turned to religion, and looks forward to the productive directions future scholarship on religion and early modern letter writing may take.

References

An account of the society in England for promoting Christian knowledge. 1817. *The Christian journal and literary register* 1: 371–7. New York: T. and J. Swords.

Arblaster, Paul. 2004. *Antwerp & the world: Richard Verstegan and the international culture of Catholic reformation*. Leuven: Leuven University Press.

Bickersteth, Edward (ed.). 1837. *The letters of the Martyrs*. London: John F. Shaw.

Brant, Clare. 2006. *Eighteenth-century letters and British culture*. Basingstoke: Palgrave Macmillan.

Constable, Giles. 1976. *Letters and letter-collections*. Turnhout: Brepols.

Daybell, James. 2005. Recent studies in sixteenth-century letters. *English Literary Renaissance* 35(2): 331–62.

Daybell, James. 2006a. *Women letter writers in Tudor England*. Oxford: Oxford University Press.

Daybell, James. 2006b. Recent studies in seventeenth-century letters. *English Literary Renaissance* 36(1): 135–70.

Daybell, James. 2009. Women's letters, literature and conscience in sixteenth-century England. *Renaissance Studies* 23(4): 516–33.

Donne, John. 1651. *Letters to several persons of honour*. London.

Gerlo, Aloïs. 1971. The *Opus De Conscribendis Epistolis* of Erasmus and the tradition of the *Ars Epistolica*. In *Classical influences on European culture A. D. 500–1500*, ed. R.R. Bolgar, 103–14. Cambridge: Cambridge University Press.

Jablonski, Daniel. 1822. An account of the people called Wenden in Germany. In *The Cambro-Briton and general Celtic repository*, vol. 3, ed. John Humffreys Parry, 477–81. London: W. Simpkin and R. Marshall.

Jackson, Ken, and Arthur F. Marotti. 2004. The turn to religion in early modern English studies. *Criticism* 46(1): 167–90.

Jacob, W.M. 1996. *Lay people and religion in the early eighteenth century*. Cambridge: Cambridge University Press.

Littlehales, Margaret Mary. 2001. *Mary Ward: Pilgrim and mystic, 1585–1645*. New York: Continuum International.

Locker-Lampson, Sophie Felicité (ed.). 1910. *A Quaker post-bag: Letters to Sir John Rodes of Barlbrough Hall ... and to John Gratton of Monyash, 1693–1742*. New York: Longmans, Green.

Marotti, Arthur F. 2000. Manuscript transmission and the Catholic Martyrdom account in early modern England. In *Print, manuscript, & performance: The changing relations of the media in early modern England*, ed. Arthur F. Marotti and Michael D. Bristol, 172–99. Columbus: Ohio State University Press.

Monaghan, E. Jennifer. 2007. *Learning to read and write in colonial America*. Amherst: University of Massachusetts Press.

Patrides, C.A. 1981. The epistolary art of the renaissance: The biblical premises. *Philological Quarterly* 60(3): 357–67.

Southwell, Robert. 1587. *An epistle of comfort to the reverend priests, and to the honourable, worshipful, and other of the lay sort, restrained in Durance for the Catholic faith*. London.

Wesley, John. 1960. *The letters of the Rev. John Wesley*, vol. 8, ed. John Telford. The Wesley Center Online http://wesley.nnu.edu/john-wesley/the-letters-of-john-wesley/wesleys-letters-1787/ Accessed 1 Nov 2011.

Whyman, Susan E. 2010. *The pen and the people: English letter writers, 1660–1800*. Oxford: Oxford University Press.

Part I
Protestant Identities

Chapter 2
Scribal Networks and Sustainers in Protestant Martyrology

Mark Greengrass

This essay examines the culture and power of epistolarity among the Protestant communities 'under the Cross' in France from 1547 to 1559, and in England from 1553 to 1558. It is based on research undertaken in the British Academy John Foxe Project, which I directed from 2004 to 2008, where we have sought to recover the English martyrologist's sources of information.[1] In the case of Foxe, we have some manuscript materials and scribal copies from which to reconstruct something of the epistolary culture and circulation upon which the protestant communities and their sustainers in Marian England relied. For his French counterpart, Jean Crespin, no such evidence now survives, although it is clear from the published editions of his martyrology that these communities also had such networks. The 'letters of the martyrs' used by Crespin and Foxe tell us something about the significance of the letter as a way of creating a faith-community vicariously, and how they cemented virtual relationships through letters in the face of persecution.

[1] For the project, see the published edition of 'The Acts and Monuments Online' (TAMO), published on 30 June 2011 at http://www.johnfoxe.org. I acknowledge the financial assistance of the Pilgrim Trust, UK and the Arts and Humanities Research Council (AHRC) for the research undertaken in connection with this paper. An enlarged version of this paper was published in collaboration with the project officer, Thomas Freeman, as 'Scribal Communication and Scribal Publication in early Calvinism: the evidence of the letters of the martyrs' in Irene Dingel and Hermann J. Selderhuis (eds), *Calvin und Calvinismus. Europäische Perspektiven*. Veröffentlichungen des Instituts für Europäische Geschichte Mainz Beihefte, Band 84 (Göttingen: Vandenhoeck and Ruprecht, 2011), pp. 394–418. Parts of the material are reworked here with the permission of that publisher. Dr Thomas Freeman undertook some of the transcriptions from Emmanuel College Cambridge manuscripts referred to; Dr Thomas Leng typed them up. Their assistance is also gratefully acknowledged.

M. Greengrass (✉)
FRIAS History, Albert-Ludwigs Universität Freiburg, Stadstr. 5, D-79000,
Freiburg im Breisgau, Germany
e-mail: m.greengrass@sheffield.ac.uk

A. Dunan-Page and C. Prunier (eds.), *Debating the Faith: Religion and Letter Writing in Great Britain, 1550-1800*, International Archives of the History of Ideas 209,
DOI 10.1007/978-94-007-5216-0_2, © Springer Science+Business Media Dordrecht 2013

Foxe and Crespin compiled their text from the materials that came into their hands (Gregory 1999, 2007). In successive editions of their martyrologies, they incorporated more and more first-hand material, engaging in a mildly competitive bid for canonical completeness. Along with depositional materials from trials, credible or eye-witness reports of their execution, the letters of the martyrs themselves were always given pride of place. That was because the latter were regarded as surviving proof that the martyrs had died knowingly and intentionally witnessing for the gospel. This was an important point, since Catholic critics of the Protestant martyrological tradition countered that they had been legitimately executed for sedition and treason, rather than executed for their true faith. Miles Coverdale, the Yorkshire Bible translator, considered the letter to be a kind of Protestant answer to the Catholic relic. Coverdale regarded it as a sort of 'techlepathy', a technology for rendering the absent present. Erasmus' extraordinary endeavours in editing St Jerome's letters reminded everyone that this was the 'thing alone that maketh men present which are absent. For among those that are absent, what is so present, as to hear and talk with those whom thou lovest?' (Pabel 2008) For Jerome, as for Erasmus, letters revealed the 'whole person'—'as in a clear glass'—the private thoughts as well as the public persona. Coverdale regarded letters as having the capacity to reveal those who suffered for the faith in a kind of three-dimensional reality:

> how the same dear children of God in their time behaved themselves … yea, what the very thoughts of their hearts were, when they prayed … when they confessed their sins and complained unto God; when they gave thanks; when they were persecuted and troubled; when they were by the hand of God visited; when they felt, not only the horror of death … but also the sweet taste of his [God's] great mercy and eternal comfort, through Jesus Christ, in their conscience.[2]

Crespin and Foxe searched for such letters from the faithful under persecution with differing success. For Crespin, the letters and confessions received in Geneva from five students from Lausanne, imprisoned in Lyon, were the stimulus to the first edition of his martyrology, published in 1554 (Truc 1920). Letters were fundamental to the way that news was manufactured; and the martyr was newsworthy for an editor like Crespin (Civil and Boillet 2005, section 3). That meant, however, having to keep the collection up to date. He duly added further letters in 1556 (Gilmont 1981a, b). By the folio edition of 1570, over 100 letters relating to the French-speaking martyrs had been included in the text, mostly written by those in prison, awaiting sentence or about to be executed. This, however, cannot have satisfied Crespin, whose location in Geneva restricted considerably the supply of material that came into his hands. The edict of Châteaubriant of June 1551 declared all those caught writing letters to Geneva, as well as anyone found carrying them, prosecutable as

[2] Miles Coverdale, *The Letters of the Martyrs: Collected and Published in 1564*, ed. Edward Bickersteth (London: John F. Shaw, 1837), xxv, thereafter LM; see also Miles Coverdale [viz: John Bull], 'The Preface to the Reader', *Certain Most Godly, Fruitful and Comfortable Letters of Such True Saintes and holy Martyrs of God…* (1564), Aiij, thereafter Bull.

though they were heretics themselves.[3] Correspondence was dangerous to its recipients, as Calvin recognized in the letter Crespin published to the students imprisoned in Lyon on 10 June 1552: 'I have put off writing to you until now, fearing that if the letters had an unfortunate encounter it would be a new excuse for the enemies to afflict you more harshly'.[4] The Parlement of Aix had already in 1548 closely interrogated a man carrying letters from former residents of Provence who had taken up exile in Geneva to their relatives—a pointer to an already quite well-organized refugee network in which relatives kept in touch and supported one another, before and after arrival in Geneva (Audisio 1999, 2005). Crespin sought to exploit such networks wherever he could, but it became harder after 1551. In Rouen, for example, two messengers from Geneva were seized in 1552. The first was French, and heading for England. The second carried letters from Normandy to Geneva. Both were executed (Monter 1999, 128). Four years later, a schoolteacher at Le Havre was captured and, having been found to be in possession of a letter from Geneva headed 'to my brothers and good friends at Havre de Grace', was similarly put to death (Monter 1999, 155–6). Elaborate aliases disguised the identities of both the individual from whom the letter was sent and to whom it was addressed in French Protestant congregations in the 1550s. They helped protect the scribal networks that were essential to the functioning reality and dynamism of emerging French Protestant Church. But as Crespin ruefully observed, 'l'ingratitude par trop vilaine' of his contemporaries, and their failure to provide him with the necessary copy, seemed to be at the root of the reason why: 'quant au nombre de ceux qui ont enduré la mort en ces derniers temps pour maintenir l'Evangile, il est notoirement tel, qu'à grand peine une petite portion en est-elle venue à nostre cognoissance' (quoted in El Kenz 1997, 129–30). Crespin did not have anything like the quantity of exemplary martyr letters that he wanted to incorporate into his martyrology.

Crespin confronted a further problem. The first exchange of letters between Calvin and the imprisoned students in Lyon from Lausanne had almost certainly been made available to him by Calvin's secretary, Charles de Jonvilliers. Jonvilliers guarded those papers carefully, planning to edit an 'authorized edition' of them in due course. That probably explains why only nine letters from Calvin to those imprisoned in France found their way into Crespin's martyrology between 1553 and 1564. Only much later—as it turned out in 1575—was a selection of Calvin's letters published in a Latin edition from Geneva, edited by Théodore de Bèze (the latter's correspondence remaining, like the registers of the Company of Pastors, a closed book to Crespin).[5] There were elaborate and, to some extent, contradictory strategies at work here, the pressures to publish being balanced by the equal and opposite

[3] Edict of Châteaubriant, June 1551, in *Recueil des anciennes lois françaises*, ed. F.A. Isambert, 29 vols (Paris, 1821–1823), 13: 203–4.

[4] Jean Calvin, *Opera quae supersunt omnia*, ed. G. Baum, E. Cunitz *et al.*, 59 vols (New York: Johnson Reprint, 1964) 14: n° 1631, thereafter CO.

[5] Jean Calvin, *Ioannis Calvini Epistolae et Responsa...* (Geneva, 1575).

concern not to provide ammunition, consciously or otherwise, to Calvin's numerous posthumous critics, not to mention the sixteenth-century equivalent of rival claims to intellectual property.

That selection of Calvin's letters already revealed how reticent he had been about encouraging the communities in France to form themselves into Churches in the early 1550s. For the historian Etienne Trocmé, the roots of what he described as the 'révolution mal-conduite', which afflicted the emerging French protestant movement, lay in Geneva (Trocmé 1959, 160–8). Calvin consciously delayed the formation of French Churches prior to 1555, only to discover that he could barely handle their explosive growth thereafter. Calvin's letters explain that reticence differently—and also serve to provide the context in which they were read. Calvin's commentaries on the Paulinian epistles had been published before 1550 and enjoyed great success. In his expositions of these texts, he invited young congregations to put themselves in the shoes of the disciples—of Timothy, Titus and Philemon, and to imagine themselves as the latter-day Galatians, Ephesians and Colossians. Paul's influence on Calvin extended well beyond the rhetorical. They influenced the contents and thought processes in them as well. Calvin mirrored Paul in his doses of fraternal encouragement served up with an equal measure of censure at the divisions in their midst, laced with warnings about the hostile world around. Calvin's deliberation was the consequence of his Paulinian model of a Christian community. He envisaged the latter slowly coalescing in mutual charity, learning how to conduct its affairs with order and decorum. If Protestants went about constructing communities too quickly, they might attract people who, although initially having 'gousté la verité de Dieu' would eventually 's'égarer à leur perdition'. It was better to proceed slowly and cautiously, with 'train et ordre', and 'convenance et accord'.[6]

Foxe's situation was different from Crespin's. He, too, understood the significance of the martyr letter and had already begun to explore its possibilities in the early Latin edition of his martyrology, the *Rerum*, published towards the end of his residence in Basel in September 1559.[7] Once returned to England, Foxe began collecting materials for the English edition, spurred on by the London printer with whom, from the autumn of 1559, he collaborated: John Day. On 10 November, Day published the first of the English martyr-letters that Foxe had begun to compile: *A frendly farewel which Master Doctor Ridley... did write*.[8] He was assisted by those in high places around the new Archbishop of Canterbury, Matthew Parker. In autumn 1560, he undertook a research trip to Norwich, and perhaps elsewhere too, looking for surviving testimonies and evidence of recent Marian martyrs. Back in London by August 1562, Foxe oversaw the printing of the *Acts and Monuments*, the first vernacular edition of Foxe's history, which appeared on 20 March of the following

---

[6] CO, 14, n° 1825 (12 Oct. 1553); 15, n° 1977 (19 June 1554).

[7] John Foxe, *Rerum in Ecclesia gestarum...* (Basel, 1559).

[8] [Nicholas Ridley], *A frendly farewel which Master Doctor Ridley, late Bishop of London did write beinge prisoner in Oxeforde, vnto all his true louers and frendes in God* (1559), Short Title Catalogue n° 21051, thereafter STC.

year from Day's print-shop. It was probably in that shop that it was decided to publish the letters relating to the martyrs as a separate publication, probably because the material had simply become too copious. The surviving manuscripts suggests that there were so many, and in different copies, that it was difficult for them to keep track of them. Foxe included a letter in the 1563 edition and attributed it to the pen of John Philpott, writing to Protestant prisoners in Newgate. Then, reflecting that he had made an error, he asked the printer to add:

> Here we must crave, gentle reader, of thee, by way of deprecation, thy favourable patience, for that in the letter above printed, for the great heap of things, and unadvised haste, we have taken one person for another, referring the letter to master Philpot, which in deede was written by John Careless.[9]

This profusion perhaps explains the publication in 1564 of the *Certain most godly fruitful, and comfortable letters of such true saintes and holy martyrs*, a year after the first English edition of the martyrology. Miles Coverdale was generally attributed as its editor, but in reality he only contributed the preface. Most of the work was compiled from copies of the letters, collated by Foxe and Bull, with occasional intervention from John Day himself (Wabuda 1993). Foxe probably knew Bull well—they had been at Magdalen College, Oxford together in the 1540s. Bull's commitment to printing these 'remains' of the martyrs appeared early on in a pamphlet letter from Bishop John Hooper from 1555, a text which answered the charge that he had encouraged Protestants to curse Queen Mary.[10] Bull and Foxe shared the goal of recovering the testimony of the martyrs 'of our times' from the 'the pit of oblivion' for 'the increase of the gospel'.[11] But their collaboration was tinged with competition too. That is something we can detect because, highly unusually for any printing in the sixteenth century, we have, among manuscripts in the British Library and at Emmanuel College, Cambridge the 'cast-offs' used in the print shop.[12]

To understand the relationship of these manuscripts to what was eventually printed is like entering someone's study and looking at the various drafts of a manuscript that is in preparation. The manuscript letters survive in various forms, and, for the supporting details, I refer to a parallel publication where I have outlined them in greater detail (Greengrass 2011).[13] We can discern four different states of survival:

[9] John Foxe, *Acts and Monuments…* (1563), 1450, thereafter A&M followed by date of the relevant edition.

[10] [John Hooper], *An Apologye Made by the Reuerende Father and Constante Martyr of Christe Iohn Hooper Late Bishop of Gloceter and Worceter Againste the Vntrue and Sclaunderous Report…* (1562), STC 13742.

[11] A&M (1563), B6r–v.

[12] British Library Additional Manuscript 19,400, thereafter BL Add. MS; Emmanuel College, Cambridge MSS 260–2, thereafter ECC. The college library's special collections are closed for extensive refurbishment (2009–2011). I have worked from microfilm copies of them.

[13] This article contains a more extensive analysis of those letters which Foxe did not publish and constitutes an earlier recension of this current piece. I am grateful to the editors for permission to reconstitute the main argument here.

(a) *Autograph* letters from martyrs or their sustainers, which Henry Bull or John Foxe, or both, gloss by adding one or more of the following in order for the manuscript to be passed to the printer: a heading, marginal glosses, deletions, additions.
(b) *Scribal* copies of letters from martyrs or their sustainers, probably produced in the print-shop and possibly from autograph originals, to which either Foxe or Bull, or both, add glosses, or otherwise edit with headings, marginalia, deletions or additions.
(c) *Fair copies* of letters from martyrs or their sustainers in Bull or Foxe's hand, often prepared with a heading and marginal glosses, perhaps from earlier copies which had become too ragged or indecipherable to pass on to the printer.
(d) *Fair copies* (including translations) from martyrs or their sustainers in an anonymous hand, and which have no sign of being edited at all.

What survives among these manuscripts must be only a fraction of what must once have existed. Only about 20% of the 206 martyr-letters published in the 1564 collection by Henry Bull is reflected in a pre-existing manuscript state. What remains, however, consists of a broad range of the martyrs in question—ten of the 22 (plus one group-letter) correspondent-martyrs leaving their mark in manuscript form. If we take what has survived as furnishing useful, albeit incomplete, clues as to the functioning of the scribal networks among the early English protestant communities, it is evident how much information Bull and Foxe suppressed. The endorsements of autograph letters, for example, provide evidence that they had actually been sent. These were not compositions in an epistolary genre, but sent as letters. The existence of several extant manuscript copies of the same letter points to the scribal networks that were in place among the Marian protestants. So, for example, there are four manuscript copies of the letter written by the martyr John Bradford on 19 November 1553 to Joan Wilkinson and the Warcups.[14] Its subject-matter—that the Marian persecution was God's punishment on the protestant reformation under Edward VI for not having completed its task, coupled with the insistence upon not attending the mass—was seen as of general significance, over and beyond the importance of its contents to the recipients of the letter.[15] One of these copies has an additional heading.[16] Another is identifiable as in the hand of William Tims, who seems to have been at the centre of a copying network operating from Newgate prison until he was martyred in 1556.[17] Another includes the telling phrase: 'Most hartely desiring you to write it owte to ye rest of *our* brethren and sisters to ye praise of *our* god. Amen.'—albeit the community in question may have been an unorthodox congregation of 'free-willers'.[18] Sometimes we pick up evidence for the couriers

[14] ECC MS 260, fols 45r–46r; 59r–62v; see also, 66r–67r; 83r–84v.

[15] Printed in Bull, 280 *et seq* [LM, 215–9]. This is an example of a letter printed initially by Bull but then followed by Foxe (1570), 1817 *et seq.*

[16] ECC MS 260, fols 45r–46r: 'Verbum domini manet in eternum'.

[17] ECC MS 260, fols 59r–62v—the copy includes an important note added by Tims, not included in any of the published version of the letter.

[18] ECC MS 260, fol. 48r.

within prison (Punt, for example, between Bradford and the free-willers in Newgate prison). These references—like the apology that John Careless included in his letter to Elizabeth Fox for having forgotten how to contact her messenger—were all silently elided from the printed versions by Foxe or Bull.[19] Indeed, it seems that Foxe and Bull deliberately sought to remove the names and to blot the traces of these networks from the published letters. They presumably did not want to compromise those who had been involved. It was still the case in the early Elizabethan régime that there were old scores to settle from the Marian divisions. Nor did they want to detract from the central point of valiant martyrs confronting their destiny, the strength of their resolve and knowledge of the truth leading the faithful.[20] Who carried the letter and how was not consonant with the image of martyrs, seeking higher truths. So Robert Glover's missive, writing to Augustine Berneher (another important intermediary in the martyr networks) would never be printed, perhaps because its reference to Mary Glover's sowing shirts for the imprisoned Latimer seemed so banal.[21] But that should have been the point. The letters of the martyrs were about the day-to-day as well as about witnessing God's truth. Not surprisingly, the editors also removed any hint of fainting resolve, or doubts. So, the poignant postscript added by Thomas Hawkes to his letter back to his native congregation is silently excised: '[signature] that shall seale thys at Cockshall [i.e. Coggeshall, the place in Essex where he was executed] ye Tewssdaye after trynyte syndaye pray pray pray even wt yowr whole hartes praye unto hym that ys abell to helpe'.[22] In a few instances we can detect the avenues by which these copies came into the hands of Foxe or Bull. The autograph letter from John Careless to Margery Cooke, for example, was evidently furnished by the Cooke family from a third party ('Upcher') since a note in Bull's hand at the top of it says: 'Mr Cooke restore this to Upcher'. There was possibly every good reason for this detail not to see the light of day because the individual in question may well have been Thomas Upchard, one of the more active 'free-willers' to emerge from obscurity, of whom more in a moment. And since we have the eventual published versions of many of these letters in various publications to emerge from Day's print-shop, sometimes in more than one edition, we can therefore begin to form some idea of the various editorial routes that this material followed in order to make it through the print-shop.

The role of the sub-editor was to mark up the cast-off in order to indicate to the type-compositors how to set it in print. Bull and Foxe were both intrusive sub-editors. They changed the paragraph structure, introduced page-breaks, catch-words and

[19] ECC MS 260, fols 144r–145v for Punt; ECC MS 260, fol. 261v for the Careless reference: 'it was ye wont off one to convey my letters unto you for I had forgot where he dwelled that was sent from yo[u]r ladyship unto me., And agayne he told me, yt he had been here to have spoken w[i]t[h] me: but ye keper wold not let hym to speake w[i]t[h] me, so that we bothe have our excuses…'.

[20] For the removal of names, see for instance BL Add MS 19,400, fols 69r–70r; ECC MS 260, fols 65r–v.

[21] ECC MS 260, fols 80r–81v.

[22] ECC MS 260, fol. 57r.

insertions into the text. They inverted word-orders, deployed synonyms, changed capitalisation, and repunctuated manuscript copies so that they lay on the page better, or read more easily. They rewrote passages where the meaning was unclear and, in at least one instance, substituted other texts for parts of letters that were missing. They added headings to them as well as marginal glosses to give additional weight and coherence to the published word. All that we might expect from a sixteenth-century editor, although we often need to be reminded that it is taking place since it (albeit unconsciously) shapes the way that we read our primary sources, especially when those are most often consulted in nineteenth-century editions which have further altered it in an attempt to render it more readable.

What is more surprising is the extent to which both Bull and Foxe altered the fundamental sense of the texts they chose to publish, and deliberately decided not to publish altogether a further significant number of letters, some of which have never seen the light of day in edited form. We have to infer the reasons for these editorial decisions. In some cases, perhaps, the correspondence was seen as too immediate, or too personal. In some instances it only existed as a fragment. We have examples, however, which suggest two very clear patterns from what was omitted or deleted. Firstly, the editors were sensitive to the already well-developed hostile trope of protestant congregations as hotbeds of promiscuity, sexual wantonness or perversion (Racaut 2002, chapters 4 and 6; Duffy 2009, 72). It was easy to deflect such criticism in advance, however, by changing the gender of the addressee of the letter, as Henry Bull did in that from John Careless to Margery Coke, replacing 'sister' with 'brother' in the body of the text, and substituting 'wife' for 'bedfellow' in another.[23] Letters in which martyrs seemed to be on too familiar terms with women sustainers, on which the latter knew and spoke their mind too forwardly, even appearing to confront patriarchal authority, were similarly silently suppressed (Freeman 2000).[24] Those relating to Sir James Hales, the Justice of Common Pleas, arrested for his Protestant views who tried to commit suicide in prison before drowning himself, were also carefully doctored to distance the martyrs from the charge that heresy drove people to madness.[25] Bull (and even more zealously, Foxe) sought to remove any suggestion of division among the ranks of the godly. In particular, they eliminated the debate about 'free-will'. The little that we know about the local groups of 'free-willers', mostly of modest means and education who had already begun to oppose the emerging orthodoxy of Edwardian England before 1553, has already been analysed by Andrew Penny (Penny 1990). They congregated mainly in south-east England, perhaps because this was where they could be in touch with the Low Countries and Calais, where there were similar dissenting groups. In Mary's reign they were incarcerated along with other protestants. So the debates continued in the King's Bench and other prisons. Only a few of the individuals concerned

[23] ECC MS 260, fols 236r–v.

[24] For instance, ECC MS 260, fols 73r–74v (John Bradford to Joyce Hales); fols 49r–50r (Anne Knyvet to John Careless).

[25] ECC MS 260, fols 34r–37r.

emerge from obscurity—Henry Hart, Thomas Upchard, John Simpson, or Robert Cooke, a clothier, Anabaptist and keeper of the royal wine-cellar (Martin 1989, chapter 6). Foxe and Bull did not want to publicize their views. So letters from suspected free-willers were cut out.[26] Others had the offending passages edited out.[27] Hart's 'Articles' of faith and the rebuttal of them by his fellow prisoners were similarly dropped.[28] The result was to fashion the Marian martyrs into a homophonous group. Something similar was at work in Crespin's narrative too, although the evidence is more tangential. We can deduce, for example, from Calvin's famous letter to the congregation of Poitiers of February 1555 that he was alert to the manuscript and oral circulation of views hostile to his own on predestination. In this case, they were stimulated by Jean Lavau de Saint-Vertunien, a disciple of Castellio in Basel.[29] With still unresolved and mounting tensions within Geneva itself (some of which had, of course, focused on this identical issue), Calvin's reticence towards an untrammeled growth of French Churches can only have been heightened. As he put it in a resigned way to the faithful in Poitiers: 'ie confesse quil nous fault porter patiemment beaucoup de faulx blasmes … aultrement il nous fauldroit tousiours avoir la plume en la main, veu que beaucoup de mesdisans ne cessent de nous denigrer tant quilz peuvent'.[30] In the successive editions of Crespin's martyrology, however, these internal tensions have been removed, and the distance between 'Rome' and 'Geneva' enlarged (Wanegffelen 1997). The martyrologists were the avatars of confessionalization 'par la lettre', using the printed evidence from scribal networks to turn early Protestantism into something more rigid than it had been.

Inprisonment was the fact of life that conditioned the writing of many of these letters, both materially and spiritually. They were letters, written by people who lacked ready access to paper, pens, writing surfaces and light. Their editors emphasize that element, seeking to access an emerging common-place in sixteenth-century society, that having a place to write was a mark of civility. 'Thus fare you well. I had no paper; I was constrayned thus to write' complains Ridley in a letter to his chaplain Augustine Berneher.[31] 'Let some bodie buy for me a pensill of lead to write with al, for I shal hardly have pen and Inke here sith all liberty of writyng is taken away from us' complains Laurence Saunders from the Marshalsea.[32] Even when they had

---

[26] ECC MS 260, fols 47r–48v [John Simpson to a scattered congregation].

[27] For instance, ECC MS 260, fols 49r–50r [Anne Knyvet to John Careless]; 130r–131v [John Careless to protestant prisoners in Newgate]; 132r–133v [John Careless to Magery Cooke]; BL Add MS 19,400, fols 76r–77v [Joyce Hales to John Careless].

[28] ECC MS 260, fols 87r–v.

[29] CO 16, n° 2118 (Feb. 1555); see also *Lettres de Jean Calvin, recueillies pour la première fois et publiées d'après les manuscrits originaux par Jules Bonnet. Lettres françaises*, 3 vols, ed. Jules Bonnet (Paris: Meyrueis, 1854) 2: 14–6. On Jean Vertunien de Lavau's connections with Basel, Castellio and Servetus, see E. Droz, 'Les étudiants français de Bâle', *Bibliothèque d'Humanisme et de la Renaissance* 20 (1958): 108–42.

[30] CO 16, n° 2118 (Feb. 1555).

[31] Bull, 73 [LM, 54].

[32] Bull, 195 [LM, 148].

pen and ink, they had to keep it secret ('for my keepers may not know that I have pen and ink' says John Bradford).[33] Some of them reported using shards of lead from the windows. For his part, Crespin reproduced the dramatic letters of the imprisoned Flemish martyr Alexandre Deykin, 'escrites de son propre sang faute d'encre…'.[34] 'Je vous eusse escrit plutost, si j'eusse eu papier & escritoire' wrote Claude Monnier, prisoner in the *officialité* of Lyon.[35] Squatting on his pallet of straw in the Bishop of London's coal cellar, John Philpott signed himself off in a letter to John Careless: 'I can write no more for lack of lyght, and that I have written I cannot read myself, and God knoweth it is written farre uneasily … written in a Colehouse of darknes, out of a pair of painful stockes…'.[36] To Robert Harrington at about the same time, he brings the letter to a close, perhaps because he was interrupted: 'Written in post haste because of strait keeping'. The scene was illustrated with a woodcut, already present in the first edition of Foxe's martyrology, in which Philpott is depicted huddled in the coalhouse clutching what might either be a manuscript or a book. Such deprivation was, however, out of the ordinary.[37] Sixteenth-century prisons were, in our terms, privatized institutions. You paid for the services that you received. Although the prison guards strip-searched those who visited prisoners, they were quite open to bribery. And in Mary's reign, the prisons where suspected heretics were detained—the King's Bench and the Marshalsea—were governed by Protestants who allowed the prisoners in their charge considerable latitude (Freeman 2004, 237–8). The fact that so many letters were despatched from protestant prisoners, many already condemned and on death row, suggests that they expected to have access to pen and paper. Several letters make it explicit that correspondence with the outside world was shared among prisoners as well as sent from one prisoner to another. It was an integral part of the judicial process that they should be expected to be able to petition the judicial authorities. They also expected to be allowed to draft a final will and testament. The authorities themselves wanted to give them the opportunity to recant. Those awaiting their martyrdom worried, of course, about incriminating the people to whom their letters were addressed. They insisted that they be sealed 'after the merchants fashion, that they be not opened'. They gave instructions to the recipients to minimize their being compromised. This, however, was part of a more general 'epistolary anxiety', the fear that letters had gone astray, been deployed for nefarious political purposes, or been otherwise interfered with (Schneider 2005, chapter 2). Foxe exaggerated for effect when he wrote that, in Mary's reign, 'almost all the prisons in England were become right Christian Scholes and Churches, so that there was no greater comfort to Christian hartes, then to come to the prisons, to behold their vertuous conuersation, and to hear theyr prayers,

[33] LM, 207; see also, 225.

[34] Jean Crespin, *Histoire des vrays tesmoins de la verité de l'Evangile…* (1570; Liège: Centre National de Recherches d'Histoire Religieuse, 1964), Book 7, fol. 611v, thereafter Crespin.

[35] Crespin, Book 3, fol. 182.

[36] Bull, 229; LM, 176.

[37] A&M (1563), 1418.

preachynges, most godly exhortations, and consolations'.[38] But he echoed what the prisoners themselves wrote. John Bradford told some newcomers to Newgate prison in London (John Hall and his wife) that they were entering God's 'Schoolehouse … that therby you might see hys carefulness and love toward you'.[39] In reality, prisons were places where mortality was high and the conditions of detention miserable. If Protestants incarcerated in France and England managed to turn their misfortune to their advantage, it was despite their predicament and often as a result of their education and training (Sherman and Sheils 2009).[40]

Rote-writing was part of school discipline. So, too, was the composing of letters. Both were part of a learned pattern of spirituality that involved private prayer, meditation and the memorising of biblical passages by heart. 'Si mon corps est enserré entre quatre murailles, l'esprit a grande occasion de se resiouir en son Dieu, puis qu'il me fait tant d'honneur de me faire compagnon de son Fils, & luy tenir compagnie à porter la croix' writes Claudes Monnier in a letter from his prison in Lyon.[41] The inflection of biblical language and metaphor in these early Protestant letters did more than give a distinctive tone and character to the letters of the martyrs (Gregory 1999, 127). It defined their mental horizons and dictated the logic of their epistolary culture. That logic included a particular emphasis that was accorded to their written word. Letters distanced themselves from it, creating an alternative and virtual epistolary community of God's elect, whom God had chosen for their particular courage and endurance.

How did the martyrs themselves regard the letters that they wrote? Four elements stand out. Firstly, they talked of them as having special significance. They were 'tokens' ('gentle token'; 'living token'; 'living remembrance'), meaning that, like the Eucharist itself in Calvinist theology, they signified the grace of God and the metonymies of a unique act. They were also 'lights', 'emblems', 'keys'; or, again, they were Chapters in the Book of Life. They offered 'comfort' to their readers who vicariously could relive their hopes and fears, and quell the rumours that were readily fostered by the authorities about their imminent recantation: 'False tongues wil not cease to lie, and mischevous hartes to imagine the worst', said Ridley to Bradford, aware that they were a small minority, engaged in a very unequal struggle.[42] They expected their letters to be read out loud within prayer congregations and therefore to share their spiritual insight about how to live 'joyfully under the cross'. That seems to have been the objective of John Hooper's letter to 'certain Godly Persons', for example, which urged them 'many tymes to have assemblies together of suche menne and women as be of your religion in Chryste' in order to 'talke and renew amonge your selues the truth of your religion … comforte one another, make prayers

[38] A&M (1570), 1696.

[39] Bull, 374; LM, 287.

[40] See especially, Thomas S. Freeman, 'The Rise of Prison Literature', 133–46.

[41] Crespin, fol. 182.

[42] Bull, 70 [LM, 51].

together, conferre one with an other'.[43] Crespin wanted to allow the martyrs to share their more intimate fears and senses of foreboding. The French protestant Bernard Seguin was worried at the 'grande fragilité qui est en nous' and how the prison authorities sought to prevent them singing Psalms together. He drew strength, however, from a new convert to the faith within the prison walls, one to whom he wrote a letter that reflected his revived sense of confidence in belonging to God's elect.[44] Others expressed their natural anxieties that their faith will be found wanting at the scaffold. The 'last farewells' of the martyrs were particularly distinctive letters of spiritual consolation, even more personal than the 'testimonies of faith' and 'confessions', generally in epistolary form, which also punctuate the letters of the martyrs. These farewell letters were necessary signs that the individuals concerned knew the cause for which they were about to sacrifice their lives. Nicholas Ridley's 'last farewells' were an emotive document, ending with a sequence of 'farewells' to the world, one that would find puritan echoes through the sixteenth century.[45] Laurence Saunders wrote a more domestic 'farewell' to his wife and friends, admitting his need for the 'comfort in my sweet Christ' to drive away 'from my phantasy the fear of death', expressing simply his faith that he will come shortly into Christ's kingdom. He offered commendations to his friends and kinsmen, 'byddyng them to beware of the Romish Antichristian religion and kyngdome, requiring and charging them to abide in the truth of Christ, which is shortlye to be sealed with the blood of their pastour…'.[46] Another urged his wife to learn from his prison experience 'by exercising your inward man in meditation of God's most holy word'. To his mother, John Bradford offered an intimate farewell letter in the form of a prayer, to which he added: 'Good mother, therefore, mark what I have written, and learne this prayer by hart, to say it daylye; and then I shall be merrye, and you shall reioyce—if that you continue, as I truste you doe, in God's true religion'.[47] In an accompanying letter, now lost, he apparently set down a further evening prayer for all the household to say, a link between martyr epistolarity and Protestant domestic spirituality which one suspects must have been somewhat in evidence, at least in Puritan circles later in the sixteenth and into the seventeenth century.

Proving that link, however, is next to impossible. It cannot be separated out from the more general impact of the martyrologies themselves, an altogether different and larger subject, still less from the impact of an evolving ambient protestant culture upon English epistolary culture. On the basis of the *Letters of the Martyrs* alone, it would be difficult to prove any case for their widespread influence upon English, and still less in the case of French Protestant epistolary practice or literary culture before 1640. The influences were set within broader parameters, some of which predated the sixteenth century, and which were determined by influences

[43] Bull, 115 [LM, 86].

[44] Crespin, fol. 214v.

[45] Bull, 80–111 [LM, 59–84].

[46] Bull, 204–5 [LM, 156–7].

[47] Bull, 453 [LM, 347].

deeply rooted in humanist and vernacular cultural traditions. In one respect, however, the martyr letters offer us a piece of suggestive literary evidence in the use (quite widespread among English Puritans) of the greeting 'Iesus Immanuel' as the header to a letter. It is to be found as a header on the manuscripts of the martyr letters (albeit eliminated in their printed versions). In some instances, these headers appear to have been added later; but it is also clear that, on others, they are contemporaneous to the copy, and sometimes in the same hand.[48] In this particular stylistic respect, if in no other, the martyr letters lay at the beginning of a broader Puritan tradition.

Secondly, letters were the way in which martyrs adopted a critical distance from the world in which they lived and would die that emerges in their heightened awareness of Satan's power and proximity that went beyond the conventional reformation polemic. For the Gascon Pierre Escrivain, writing for the scholars imprisoned in Lyon, they were in mortal combat with Satan.[49] Nicholas Ridley's Satan was a cosmic force, capable of driving down 'the third part of the stars in heaven'.[50] For Laurence Saunders, Satan was the small voices or 'frayebugges' (spectres) within.[51] To John Bradford, Satan was present throughout the world, standing 'now at every Inne door in hys city and countrye of this worlde, crying unto us to tarye and lodge in thys or that place, till the stormes be overpast'.[52] Sometimes this 'distanciation' is presented in apocalyptic terms. Martial Alba, a student from Montauban and the eldest of the students captured in Lyon in 1555 wrote to his companion 'freres fideles estans en la ville de Bordeaux' seeking to console them with a vision of the Angel of the Apocalypse coming to their assistance.[53] For John Careless too, the Apocalypse was not far from his mind when he came to say his farewells. 'Let them remember' he told a friend by letter, referring to those whose courage left them 'that in the Apocalypse the fearful be excluded the kingdom'.

For many martyrs, the world looked different when viewed from their prison. It was a matter of being 'in the world, but not of the world'. John Bradford reminded Sir James Hales, then in the Counter prison and suffering from depression: 'Let the worldlings waye thinges and loke upon the affaires of men with their worldly and corporall eyes ... but let us look on things with other manner of eyes ... you then beheld things not as a man, but as a man of God...'.[54] To Humphrey Hales, he explained how being imprisoned had radically changed his view of the world. The latter was a 'smoke, a shadow, a vapour', the 'glory of this life' was (quoting Job) no more than 'grass, hay'.[55] The knowledge of the Lord, he told a worshipping

[48] For instance, ECC MS 260 fols 11r, 40r, 72r; BL Add MSS 19,400, fols 66r, 76r, 78r, 84r.

[49] Crespin, fols 201–2r.

[50] Bull, 197 [LM, 89].

[51] Bull, 205 [LM, 157].

[52] Bull, 282 [LM, 217].

[53] Crespin, fols 198r–v.

[54] Bull, 287 [LM, 221].

[55] Bull, 311 [LM, 238].

congregation in a letter 'standeth not in forked caps, tippets, shaven crowns or such other baggage and antichristian pelf; but in suffering for the Lord's sake. The world shall hate you, saith Christ. Lo, there is the cognizance and badge of God's children'.[56] For Pierre Escrivain, too, there was a contrast between the 'ondes & vagues de la mer de ce monde' and the glory and consolation which will be their heavenly reward.[57]

Accompanying that greater critical distance upon the world, was a complementary proximation of the self. The letters were 'tokens' quite simply because the individuals writing them felt metonymously at one with themselves and with Christ. The relationship between the body and the letter was a physical one; of hand, eye and mind. There was a physiognomy of the martyr letter—the 'spirituality' of martyrdom rendered 'material'. It was an artifact that was a 'window on the soul'. So many of their letters reflected this sense of a new-found unity of soul and body, an awareness of their physical as well as their mental state and the relationship between them being a constituent element of the letters. 'Let us not … divide ourselves, and say our soules serve hym, whatsoever our bodies do to the contrarye for civile order and policie' advised John Hooper.[58] They derived from that sense of wholeness their widespread and reiterated fear of backsliding and conformism among their supporters. 'Beware of this folyshe and disceytfull collusion' wrote John Hooper on 14 June 1554 to an unidentified community, 'to thynke a man may serve God in spirite secretly to his conscience althoughe outwardly with hys bodye and bodilye presence he cleave, for civyl order, to such rytes and ceremonies as now be used contrary to God and hys word'.[59] Such nicodemism was the sign that your 'hearts are wedded to the perishing treasures of this world'. The letters of the martyrs were 'true tokens' because of the close alignment of soul and body that they proclaimed. Laurence Saunders earnestly urged the 'professors of the gospel' of Lichfield to whom he dedicated one of his farewell letters from the Marshalsea on 17 October 1554, not to 'addict yourselves unto the fantasing of the flesh-pots of Egypt' (Exodus, 16:3), surrendering the 'promised possession' of Christ.[60] There were letters to 'backsliders' and 'recanters', letters of exhortation and practical advice. John Bradford composed his lengthy treatise on the 'hurt of hearing mass' whilst in prison. His objective was to persuade his readers that the mass was a negation of Christian community, a 'plain mark of antichrist's catholic synagogue'. He reserved his greatest scorn in his prison letters for those who 'will have Christ, but none of his cross … they will be counted to live godly in Christ, but yet they will suffer no persecution … they love God in their lips, but in their hearts, yea, and in their deeds, deny him'.[61] The challenge for those reading the martyr letters, and the essence of their impact, was in a radical reordering of oneself in respect of the world.

[56] Bull, 319 [LM, 245].

[57] Crespin, fol. 202v.

[58] Bull, 119 [LM, 89].

[59] Bull, 126 [LM, 95].

[60] Bull, 184 [LM, 140].

[61] Bull, 381 [LM, 293].

This reordered self was an essential precondition for entering into the virtual community of the godly. At the heart of that community was communion with Christ. Laurence Saunders wrote of being 'graffed in Christe', 'greaffed in him as braunches in that so heavenlyes vine … knit unto him as the sundry members of that body', a 'communion and fellowship' which 'passeth all understanding'. Some martyrs spoke of the unique friendship that had been nurtured by the events that had overtaken them. Denis Peloquin explained to his nephew in a letter of 23 August 1553 that it was a friendship that was not of this world.[62] The martyrs' letters provided a representation for the invisible community of the saints to which they felt that they belonged (Greengrass 2001). That representation went beyond a rhetorical construction since their letters evidently served as a personal medium for the conveyance of strong, affective emotions of support to a wider community of believers. They addressed their letters to their 'faithful friends', 'sisters', 'relievers', 'brethren dispersed abroad in sundry prisons', 'brethren which constantly cleave unto Christ'. They encouraged their letters to be read out, and addressed them to whole communities, to be part of them, albeit virtually and vicariously. The French prisoner Pierre Bergier instructed his wife:

> afin que nos enfans ayent perpetuelle memoire de moy, ie vous prie recouurer toutes les letters que ie vous ay enuoyees, & plusieurs autres qui m'ont esté escrites, de quelles il y en a vn grand nombre par-deça chez mon beau-frere, & les faire rescrire à mon frère Denis, ou a quelque autre dans vn liure exprez. Et après les auoir fait escrire, vous pourrez distribuer lesdites lettres aux vns & aux autres…'[63]

Such re-circulation was an essential part, of course, of 'scribal publication' (Love 1993). In the case of the martyrs, however, it was emphasized by their sense that God had called them to be spiritual leaders. Their sense of distanciation from the world stimulated their desire to engage *with* it through letter-treatises of advice and counsel.

Through these prison letters we glimpse the family, kin, friendship communities, local congregations and affective relationships which sustained the martyrs. The most direct and appealing letters were those that they composed to their families, at once moral and social; recalling the debts of the past and the ties of family even as they appealed to a new sense of community. We might call these 'private' letters were it not evident that they were not simply dyadic communications, limited to the martyr and his wife, mother, or close kin, but triadic conversations, in which there was a third party: the communion with Christ and the professors of the gospel. That third party was typically evoked when it came to expressing the affective passions, especially patience and joy. Writing to his wife Ann, John Hooper composed a whole 'exhortation to patience' in which he rescripted the female virtue of patience into a foundation-stone of Christian faith.[64] Despatching a letter to his wife Jeanne

[62] Crespin, fol. 247v.

[63] Crespin, fols 236r–v.

[64] LM, 112–20.

on 7 July 1553, Denis Peloquin thanked her for his prayers and consolatory letters, commending her great patience.[65] Laurence Saunders, writing to his wife, counseled joy: 'be merry in God, in whom also I am very merry and joyful'. This would probably be his last letter, he told her: 'The keeper saith he must needs see that we write not at all'.[66] She was not to worry, though: 'The devil roareth; but be of good cheer; he will shortly be trodden under foot, and the rather by the blood of martyrs'.

References

Audisio, Gabriel. 1999. Huguenots des villes et huguenots des champs: Les premiers réformés de Provence (1530–1560). *Provence Historique* 49: 563–75.

Audisio, Gabriel. 2005. The first Provençal Refugees in Geneva (1545–1571). *French History* 19: 385–400.

Civil, Pierre, and Danielle Boillet. 2005. *L'Actualité et sa mise en écriture aux XVe-XVIe et XVIIe siècles*. Paris: Presses de la Sorbonne Nouvelle.

Duffy, Eamon. 2009. *Fires of faith: Catholic England under Mary Tudor*. New Haven/London: Yale University Press.

El Kenz, David. 1997. *Les Bûchers du roi: La culture protestante des martyrs (1523–1572)*. Seyssel: Champ Vallon.

Freeman, Thomas S. 2000. "The good ministrye of Godlye and Vertuouse women": The Elizabethan martyrologists and the female supporters of the Marian martyrs. *Journal of British Studies* 39: 8–33.

Freeman, Thomas S. 2004. Publish and Perish: The scribal culture of the Marian martyrs. In *The uses of script and print, 1300–1700*, ed. Julia Crick and Alexandra Walsham. Cambridge: Cambridge University Press.

Gilmont, Jean-François. 1981a. *Bibliographie des éditions de Jean Crespin 1550–1572*, 2 vols. Verviers: P. M. Gason.

Gilmont, Jean-François. 1981b. *Jean Crespin. Un éditeur réformé du XVIe siècle*, Travaux d'Humanisme et Renaissance, n° 186. Geneva: Droz.

Greengrass, Mark. 2001. Informal networks in sixteenth-century French Protestantism. In *Society and culture in the Huguenot world*, ed. Raymond A. Mentzer and Andrew Spicer, 78–97. Cambridge: Cambridge University Press.

Greengrass, Mark. 2011. 'Scribal communication and scribal publication in early calvinism. The evidence of the letters of the Martyrs'. In *Calvin Und Calvinismus: Europäische Perspektiven*, ed. Irene Dingel and Herman J. Selderhuis. Göttingen: Vandenhoeck and Ruprecht.

Gregory, Brad S. 1999. *Salvation at stake: Christian martyrdom in Early Modern Europe*. New Haven/London: Yale University Press.

Gregory, Brad S. 2007. Persecutions and martyrdom. In *The Cambridge history of Christianity*, Reformation and expansion, 1500–1650, vol. 6, ed. R. Po-Chia Hsia, 261–288. Cambridge: Cambridge University Press.

Love, Harold. 1993. *Scribal publication in seventeenth-century England*. Oxford: Clarendon.

Martin, Joseph Walford. 1989. *Religious radicals in Tudor England*. London: Hambledon Continuum.

Monter, William. 1999. *Judging the French reformation: Heresy trials by sixteenth-century parlements*. Cambridge, MA: Harvard University Press.

[65] Crespin, fols 250r–v.

[66] LM, 149–50.

Pabel, Hilmar M. 2008. *Herculean labours: Erasmus and the editing of St Jerome's letters in the Renaissance*. Leiden: Brill.

Penny, Andrew D. 1990. *Freewill or predestination: The battle over saving grace in Mid-Tudor England*. Woodbridge: Boydell Press.

Racaut, Luc. 2002. *Hatred in print. Catholic propaganda and protestant identity during the French wars of religion*. Aldershot: Ashgate.

Schneider, Gary. 2005. *The culture of epistolarity: Vernacular letters and letter writing in early modern England*. Newark: University of Delaware Press.

Sherman, William H., and William J. Sheils (eds). 2009. *Prison writings in early-modern England. Huntington Library Quarterly* 2: 72.

Trocmé, Étienne. 1959. Une Révolution mal conduite: A propos d'un centenaire et d'un livre. *Revue d'histoire et de philosophie religieuses* 2: 160–68.

Truc, G. 1920. Calvin et les cinq prisonniers de Lyon. *Revue des études historiques* 86: 43–54.

Wabuda, Susan. 1993. Henry Bull, Miles Coverdale and the making of Foxe's *Book of Martyrs*. In *Martyrs and martyrology*, Studies in Church History, vol. 30, ed. Diana Wood, 245–58. Oxford: Blackwell.

Wanegffelen, Thierry. 1997. *Ni Rome ni Genève: Des fidèles entre deux chaires en France au XVIe siècle*. Paris: Honoré Champion.

Chapter 3
Thomas Browne, the Quakers, and a Letter from a Judicious Friend

Reid Barbour

As much as, perhaps even more than, any other religious group in seventeenth-century England, the Quakers embraced the Pauline tradition of letter-writing as the chief literary means of communicating with fellow 'Friends' but also of reaching out to potential converts among the mass of human beings, all of whom were gifted with the divine light of Christ within them but whose sinfulness impeded their attention to that light. Of the countless letters of comfort, prophecy, instruction, and invitation sent out by Quakers in the 1650s and 1660s, one was addressed to Thomas Browne, the famous physician of Norwich and author of *Religio Medici* and *Pseudodoxia Epidemica*. Written to Browne by a well-known Norwich Quaker named Samuel Duncon in what I will argue was the period during or just after 1659, the brief manuscript document reads in its entirety as follows:

> ESTEEMED FRIEND, Haveinge perused a booke of thyne called Religio Medici (and findeinge these sound assertions followinge—'To aske whare heauen is, is to demand whare the presence of God is'—'Moyses committed a gross absurditye when with these eyes of fflesh he desired to see God.' Wee are much contested agst by some, because we can't comply to their tenett in that particular, viz. that with their ffleshly eyes they shall see God. 'There is a piece of divinitye in us, some thinge that was before the elements'—'That God loves us for that part which is, as it were himselfe, and the traduction of his holy spirit.') Judgeinge thee juditious, I therewith send thee a booke to peruse; and if thou desire any personall conferrance with me, or any of my friends concernynge the principalls of our religion, (which we believe is the immortal religion, though generally accounted herisie) I shall indeauer it, in the same loue I present this booke to thy vieue, who am a lover of mankind in generall, and thyselfe in particuler.[1]

---

[1] The letter can be found in Bodleian Rawlinson MS 391, fol. 52r; printed in *Sir Thomas Browne's works, including his life and correspondence*, ed. Simon Wilkin, 4 vols (London: William Pickering, 1836), 1.352, from which I quote here. In the manuscript a small portion of the letter is inserted in the left margin.

R. Barbour (✉)
Department of English, University of North Carolina, CB 3520, Chapel Hill, NC 27599, USA
e-mail: rbarbour@email.unc.edu

A. Dunan-Page and C. Prunier (eds.), *Debating the Faith: Religion and Letter Writing in Great Britain, 1550-1800*, International Archives of the History of Ideas 209,
DOI 10.1007/978-94-007-5216-0_3,

Despite Browne's prominent (if complex) pronouncements of his commitment to the Church of England, and despite *Religio Medici*'s disregard for 'misguided zeal',[2] Duncon looks past what he would consider the carnality and formality of Browne's work and finds three seeds of spiritual wisdom that have convinced Duncon that Browne is ready to be a friend to the Friends. What specifically has Duncon singled out from the *Religio*'s complex meditations? What do we know about Duncon? How if at all did Browne respond to this invitation? And how might the very form of what Duncon only calls 'a book of thyne' have played a part in the Quaker's expectation that Browne would welcome a letter. What, finally, does Duncon mean by referring to Browne as 'judicious', especially in the context of concerns about heresy?

One of the quotations selected from *Religio Medici*, that 'Moyses committed a gross absurditye when with these eyes of fflesh he desired to see God'—is in some measure the least to be expected from a Quaker but also the boldest restatement of an essential Quaker position. Quakers privileged the Spirit, Light, and Word of God over the words to be found, in whatever human language, in the Scriptures, though most insisted that this priority did not subvert respect for the Bible but simply prescribed how—by way of God's Spirit—the true Christian should approach it.[3] It is true that some Quaker authors stressed the human imperfections of the Scriptures, whilst others emphasized the allegorical and symbolic mysteries contained therein, and still others urged a cautious attitude that would neither 'despise' nor 'doat' on the Scriptures.[4] Whatever their differences, however, in by far the most Quaker texts published in the decade or so after the Friends first declared themselves an entity in the early 1650s, Moses is virtually always treated with considerable respect.[5] Indeed he comes across as the first of the Friends, blessed with the Spirit that dictated Scripture (as against having Scripture guide the Spirit), so filled with that Spirit as to commune with God face-to-face until he trembled or 'quaked', and staunchly resistant to his people's tendency toward idolatry and carnality. Yet implicit in the Quaker privileging of the spirit over the written words of the Scripture was a skepticism about the reliability of Scripture that has led scholars to link Quaker ideas to those of Spinoza and further to the Enlightenment and that Duncon has identified in Browne (McDowell 2003, 136–82). Indeed, it was Browne's willingness to criticize the believability of passages in Scripture that led his earliest opponents to judge him heretical and even atheistic—this despite Browne's declarations of admiration for and faith in Scripture.

Duncon has truncated Browne's own version of the Mosaic absurdity, in which the Biblical error is attributed to Moses's Egyptian learning and its seriousness somewhat limited by being relegated to the realm of 'Philosophy' rather than 'divinity'.[6]

[2] For convenience, unless otherwise noted I quote the text of *Religio Medici* from *Sir Thomas Browne: The major works*, ed. C. A. Patrides (New York: Penguin, 1977), here at 63.

[3] For the Quaker view of the Spirit in relation to the Bible, see Nuttall, 1992.

[4] For a sharper critique, see Henry Clark, *A cloud of witnesses* (1656), 3–4. For a more cautious approach, see John Toldervy, *The foot out of the snare* (1655), A3r.

[5] For a few examples, see Isaac Pennington, *Where is the wise?* (1660), 7; James Naylor, *A discovery of faith* (1653), 5; and Naylor, *An answer to twenty-eight queries* (1655), 18.

[6] *Sir Thomas Browne: The major works*, ed. C. A. Patrides, 122.

Nonetheless, Duncon's recognition of Browne's skepticism regarding the Scriptures is closely connected to the letter-writer's claims that Browne is a man with profound spiritual understanding. And so Browne comprehends that heaven and the divine are within us, now and for all time, as surely as God, the Word, Light, and Spirit must be seen with the spiritual rather than the fleshly eye.[7] Indeed, in the paragraphs following the Moses reference, Browne stresses that heaven takes up residence in the soul or, alternatively, that heaven coincides with divine presence.[8] Yet even if the basic sentiments that Duncon finds in Browne are everywhere to be found in Quaker writings, the wording of the 'piece' of God contained within us is strikingly bold and arguably unusual for at least the earliest Quakers. While early Quakers tend to insist that the kingdom of God is in us,[9] relatively few push the point to the radical conclusion that there is no such external place as heaven or hell, or insist on making God precisely a 'piece' or a 'part' of us rather than simply 'in' us. For the most part, as Rosemary Moore has explained, the Quakers erred on the side of caution in defining, as against simply heralding, their unity with the divine or the metaphysics of immanence, even if they are emphatic on how the light or the seed within us is fundamentally different from any natural faculty or capacity (Moore 2000, 80–91). Especially among the earlier Quakers in the 1650s, speculative theology was a distant concern when considered next to the apocalyptic and evangelical urgency of the Quaker mission to save humanity from its darkness (Reay 1985, 33). The learned Friend named Samuel Fisher writes relatively much and academically on heaven, the nature of the soul, and eschatology, yet he tends to fudge the ontological status of heaven by speaking of it as both 'above' and 'within' us.[10] Whereas the Quaker metaphor of the light is usefully vague for the question of the extent to which the divine participates in the human or the human in the divine, enemies of the Quakers enlist the language of the 'part' in their attacks, for instance, Francis Higginson in *A Brief Relation of the Irreligion of the Northern Quakers* (1653); indeed, Quaker self-defense includes a refusal metaphysically to have the light pinned down, as it were.[11]

---

[7] Quaker writings are so filled with these ideas as to make a choice of examples arbitrary, but see for instance James Naylor, *A discovery of the first wisdom from beneath, and the second wisdom from above* (1656), 4, 11; Richard Hodden, *The one good way of God* (1661), 12–3; Thomas Forster, *A guide to the blind* (1659), 38, 49; John Crook, *A defence of the true church called Quakers* (1659), 26–9; and Francis Howgill, *Some of the mysteries of God's Kingdom* (1658), in *Memoirs of Francis Howgill, with extracts from his writings*, ed. James Backhouse (York, 1828), 99, 105. For the vanity of imagining heaven as a distant place, see Nuttall, 1992, 135–6; on the internalizing tendency, see Reay, 1984, 147.

[8] *Sir Thomas Browne: The major works*, ed. C. A. Patrides, 125.

[9] For instance, James Naylor, *The Lambs Warre* (1657), 6–7; Naylor, *The power and glory of the Lord shining out of the north* (1656), 1; Martin Mason, *A loving invitation* (1660), 4.

[10] Samuel Fisher, *Apokrypta Apokalypta* (1661), in *Early quaker writings, 1650–1700*, ed. Hugh Barbour and Arthur O. Roberts (Grand Rapids, MI: William B. Eerdmans, 1973), 312.

[11] See Francis Higginson in *A brief relation of the irreligion of the northern Quakers* (1653) 5, point 9. For the refusal of metaphysics, see Fox and Naylor, *Saul's Errand to Damascus* (1654), in *Early quaker writings*, ed. Barbour and Roberts, 253.

Indeed, Duncon has overlooked the metaphysically safer, if spiritually more fervent, Brownean celebrations of the Spirit inherent in us all; in one instance, Browne declares his certainty that 'there is a common Spirit that playes within us, yet makes no part of us, and that is the Spirit of God'[12]; the image of play gives way to metaphors of 'fire and scintillation', of 'life and radically heat', to the Biblical creativity of 'the gentle heat that brooded on the waters, and in 6 days hatched the world', then finally from play, fire, and doves to wind, to 'the warme gale and gentle ventilation of this Spirit'. Duncon has selected Browne's most theologically daring formulation of the human merger with the divine, a question over which Browne's contemporaries from the Cambridge Platonists to the Church of England's own Henry Hammond had struggled in dialogue with the ambiguities of ancient pagan schools such as Stoicism.[13] Presumably it is in the meetings to follow that Duncon or his Friends would see fit to elaborate on specifically Quaker arguments about the human participation in the divine such as the elision between justification and sanctification, that is, between salvation and purification. And it is this anticipation of conversation that complicates the audience of the letter. Duncon has no apparent designs on publication for his letter to Browne: its terms are clearly meant to awaken Browne to his extraordinary kinship with the Quakers but also perhaps to notify Browne that he finally isn't on the side of the orthodox formalists, that he is at heart what those formalists would misjudge a heretic. Even so, it is entirely probable that Duncon would have imagined secondary audiences for his letter among Browne's family, friends, and extensive social network in Norfolk. No doubt this is part of the reason why Duncon stresses the universality of Quaker redemption as well as the pervasive misconception of Quaker heresy.

To this end, Duncon appeals to an essential impartiality purveyed by the persona of *Religio Medici*: the claim that its author tests all positions on his own, that he is therefore unbiased and not to be misled by contemporary constructions of heresy, indeed that he himself has frequently considered unorthodox ideas and, even if he has moved beyond some of those as error, he is prepared to credit men of singular visions whose wisdom captures truths either long forgotten in the Church or never before properly grasped. What is more, Browne portrays himself as opposed to partisan hostility whether it is directed toward other religions, various Christian styles of worship, or a national cuisine. At one point in *Religio Medici*, Browne insists that 'Persecution is a bad and indirect way to plant Religion; It hath beene the unhappy method of angry devotions, not onely to confirme honest Religion, but wicked Heresies, and extravagant opinions'; and persecution was at the foundation of the Christian faith itself.[14] By contrast, not only were the Quakers constantly assaulted in person and in print, but according to Richard Hubberthorn, Norwich was especially 'wrangling, mischievous, envious, malicious' against witnesses

[12] *Sir Thomas Browne: The major works*, ed. C. A. Patrides, 99–100.

[13] For the difference between Henry More the Cambridge Platonist and the Quaker position on the Light within, see Nuttall, 1992, 18; for Hammond and the Stoics, see Barbour, 1998, chapter 5.

[14] *Sir Thomas Browne: The major works*, ed. C. A. Patrides, 93.

to righteousness.[15] It is the sense of Browne's unprejudiced open-mindedness that Duncon captures with his word 'judicious', for this is the term that Quakers frequently use for those members of their community open to careful, sober consideration of their message—this, in response to the appropriation of the same word by those enemies of the Friends seeking to lampoon their zany appeals to the Holy Spirit from the vantage of a scripturally orthodox and soberly rational certainty.[16] Browne uses 'judicious' to characterize his own mode of theological inquiry.[17]

A superficially related but wholly different term, also used by Duncon, 'judging' is more problematic among the Friends themselves. Judgment was at once central to the Quaker enterprise of calling the whole world to the Spirit and of prophecying apocalyptic woe to those who remain in darkness, yet throughout their early history a source of divisiveness amongst the Friends themselves against which their leaders were constantly warning.[18] In his formulation 'judging thee judicious', Duncon encapsulates the central problem of a movement whose self-identity was at once severely exclusionary and comprehensively open to evangelical hope for the entire human race. On the one hand, then, the recalcitrant citizens of Norwich scarcely promised a few to be gathered 'as the gleanings of the vintage'.[19] On the other, Quakers warned one another to 'Beware of setting up the bounds of your love to the brethren under any form, or opinion whatsoever, but let your love be equally unto all, where you do see the appearance of Christ in Spirit and power from on high'.[20] Even if such loving rhetoric terminated with the 'true Saints', it tapped into the notion that God's spirit can be found in us all. In Browne's own usage 'judgment' can name the rigorous application of an inwardly located standard to all religious beliefs and practices.[21] But Browne also enlists the phrase 'sober judgement' as the guarantor of the boldest, most irregular thought, according to

---

[15] See Eddington, 1932, 14. On the difficulties faced by the early Quakers in Norwich and Norfolk, see Vann, 1969, 16–19; and Braithwaite, 1955, 162–5. The constant ridicule of the Quakers in Norwich is illustrated by one the first local converts to the Friends, Thomas Symonds, who had visited Quaker prisoners in order to ridicule them. See Platt, 1926, 1.

[16] For 'judicious' in the battle over Quakerism, see Francis Howgill, *The heart of New-England hardned* (1659), 12; Howgill, *The mouth of the pit stopped* (1659), 12; Thomas Ellwood, *An answer to George Keith's narrative* (1696), 181; Elizabeth Bathhurst, *Truth's vindication* (1679), 88; Robert Barclay, *William Michel unmasqued* (1672), preface to reader, 7, 29, 52; Barclay, *Quakerism confirmed* (1676), 3, 8, 12, 72, 84, 88; Francis Higginson, *A brief relation of the irreligion of the northern Quakers* (1653), 35, 65, 78; John Gauden, *A discourse concerning publick oaths, and the lawfulness of swearing in judicial proceedings* (1662), 3, 32; John Brown, *Quakerisme the path-way to paganism* (1678), epistle to reader, 2, 6, 56, 279; and William Allen, *The danger of enthusiasm* (1674), 106.

[17] *Sir Thomas Browne: The major works*, ed. C. A. Patrides, 75.

[18] See Naylor, *The Lambs Warre*, 2–4, George Fox, 'To Friends, concerning judging', in *George Fox speaks for himself*, ed. Hugh McGregor Ross (York: William Sessions, 1991), 59–60.

[19] The words of Richard Hubberthorne, quoted in Eddington, 1932, 10, from Swarthmore MSS, iv.6. For the gleaning metaphor and others like it, see Bauman, 1983, 63–70.

[20] Edward Punch, *A cryer in the wildernesse of England* (1653), A3r.

[21] *Sir Thomas Browne: The major works*, ed. C. A. Patrides, 61.

which there are 'many things untouch'd, unimagin'd, wherein the libertie of an honest reason may play and expatiate with security and farre without the circle of an heresie'.[22] In Duncon's response, then, Browne is at once the most unbiased and welcoming of readers, as well as the most spiritually rigorous and inwardly exploratory, one whose protestations of following in the circular path of the Church are overwhelmed by two rival circle images: one in which his faith is centered inside the circle of the self, and the other in which it freely plays outside the confining circumference of human forms.

Two other features of Duncon's invitation establish further sympathy between Browne and the Quakers. One is that *Religio Medici* fervently celebrates the spiritual profundity of true friendship, even at the risk of showing disrespect for parents and family.[23] More subtle, however, are the signs that *Religio Medici* was originally composed as a letter to a friend, as Browne suggests in his preface to the reader, in the work's opening gambit, and in the conclusive expression of Browne's desire for 'the love of thy selfe and my dearest friends'.[24] As with the great Biblical model of the letter writer Paul, so too with letter writers among the Friends: Quakers sent letters either in an effort to solidify the identity of the already demarcated, but geographically and even doctrinally disparate group, or as invitations to those who like Browne might not yet be aware that they belong in what Francis Howgill called the 'net' into which the Kingdom of Heaven was gathering all those who wished to be included.[25] According to a Friend called John Toldervy, such a person, far from being a scoffer, is 'One, whose Minde is (in some measure) fitted for Information; and not of those, who are Resolved into Self, and Hardened against any Discovery which might effect to the enlightening of [his or her] Understanding'.[26] And sometimes in the hands of a Friend such as Margaret Fell, the rhetoric of human unity was explicitly extended, as it is in different terms and on other foundations in *Religio Medici*, to people outside Christendom, for instance, to 'Jews and Gentiles, Barbarian, Scithian, Bond and Free, in the Unity of the Spirit which is in Christ Jesus our Lord'.[27]

Duncon's letter to Browne illustrates a dynamic and permeable—if also an uncommonly contentious—relationship between the Friends and the world that they would lead out of woe into redemption. Duncon himself appears to have undergone a conversion to the light and the spirit of the Friends in 1659. In that year, he

[22] *Sir Thomas Browne: The major works*, ed. C. A. Patrides, 69.

[23] *Sir Thomas Browne: The major works*, ed. C. A. Patrides, 59, 142–3, 160.

[24] *Sir Thomas Browne: The major works*, ed. C. A. Patrides, 59–60, 160.

[25] For Howgill, see Brockbank, 1929, 57. Scholarship on the Quaker use of letters includes Peters, 1995, 12; and Horn, 2008. For the disparate tendencies of Quaker thought, see Smith, 1995.

[26] Toldervy, *The foot out of the snare*, A2r.

[27] Margaret Fell, *A true testimony from the people of God* (1660), A2r. Cf. George Fox's letter in *'The power of the Lord is over all': The pastoral letters of George Fox*, ed. T. Canby Jones (Richmond, Ind.: Friends United Press, 1989), 171. The emphasis on outreach to the 'other' is the reverse side of the Quaker resistance to spiritual pollution, for which see Davies, 2000, 3, 36, in contrast with the Quaker program of outreach treated on 16, 108–14.

published *Several Proposals offered (by a friend to peace and truth) to the serious consideration to the keepers of the liberties of the people of England.* Prior to 1659, he was known to be sympathetic and helpful to, yet not a member of, the Quakers, but in the 1659 publication, one sees evidence of the conversion in the postscript that appears to have been hastily added to what is otherwise a practical set of recommendations regarding government funding. Suddenly in the postscript, Duncon's language drastically shifts to the prophetic register of warning and woe so familiar in Quaker writings from the 1650s, and even though he continues to distribute political responsibility among MPs, Independents, Baptists, and Quakers, his appeals to 'the Spirit and movings of God' corroborate the language of prophetic admonition symptomatic of a man who has been convinced by the Friends. Norwich records tell us, moreover, that in 1660 Duncon found himself in prison for not swearing oaths.

1659 was a wildly successful year for the Quakers, not just generally but specifically in Browne's Norwich, which prior to that year had proved difficult ground for the Friends to cultivate. But it was also the year in which the pervasive and intense fear of Quakerism—as witchcraft, mania, and anarchy—reached its peak (Reay 1985, 81–100). Tension over the Quakers often impinged quite closely on Browne's neighborhood in Norwich. Earlier in the 1650s, one prominent confrontation took place at Browne's own church in the market, St Peter Mancroft, between leading Quaker George Whitehead and the parish minister John Boatman, whilst the market itself, the site of Browne's home, was a frequent location of Quaker evangelizing.[28] Yet with the Restoration, soon after which the Quakers were singled out for oppressive legislation, the Friends were so maltreated that their emphases and tone drastically changed, becoming more passive and pacifistic, more accommodating and apologetic, less apocalyptic and more inwardly divided about mundane questions such as organization (Ingle 1994, 190). Duncon's letter to Browne is still heady in its hope that the famous citizen of Norwich will join sides with the Friends, but is also fully aware that the often anti-intellectual and egalitarian Friends need learned and prominent support, and that the bulk of contemporary East Anglian society is very much against them. Nor is the Quaker outreach to Browne impeded by his medical vocation or scientific work: as George Fox's 'Book of Miracles' and Quaker correspondence purvey, the Friends were at times resistant to recourse to medicine, but by and large were respectful of its assistances to human life—and this was the case, whether or not they committed themselves to the legitimizing proof of miraculous cures or whether they were indebted to the continental mystical writings of Boehme with his attention to the signatures of the natural world, though as we will see for one Quaker the connection between Boehme and Browne was a crucial one.[29]

[28] For the Quakers in Norwich 1659, see Eddington, 1932, 26–7; cf. *The journal of George Fox*, ed. Nigel Smith (New York: Penguin, 1998), 270–1. For preaching and prophecying in the market, see Richard Hubberthorn, *The testimony of the everlasting gospel witnessed through suffering* (1654), 4; and see also George Whitehead, *The Christian progress of that ancient servant and minister of Jesus Christ* (1725), 34. For Whitehead and Boatman, see *The Christian progress*, 34.

[29] On Fox and medicine, see *George Fox's 'book of miracles'*, ed. Henry J. Cadbury (Cambridge: Cambridge University Press, 1948).

As Duncon's letter makes plain in its conclusion, what he and his friends lovingly seek from Browne is the sharing of discourse, first by way of an exchange of books but ultimately by way of conversation. Such a love of 'free Conversation and Fellowship of sober, faithful Friends' is celebrated in George Whitehead's autobiography, in which Whitehead, a key Quaker in Browne's Norwich, connects the spirit of conversation amongst the 'sober, studious Scholars' of his early education to that of the Quakers, both groups resistant to 'loose, extravagant' company.[30] As Sarah Hutton has shown, Cambridge Platonist Henry More may not have liked the fact that his pupil Anne Conway came to welcome the 'deluded, melancholy' Quakers into her home at Ragley Hall, but articulating his disregard for these 'fanatics', at least in the 1670s, was no simple matter; having met several, More had to admit that 'There are some things which I hugely like in the Quakers' whose conversation comforted Lady Conway in severe affliction (Hutton 2004, 178, 188). With really no way to know about Browne's most explicit contribution to epistolary literature—*Letter to a Friend* was not published until 1690—Duncon does not say whether he is aware of Browne's *Pseudodoxia Epidemica*, in which Browne invites full-scale and widespread collaboration from his readers in the Baconian advancement of learning, though by 1659 it was commonplace for readers to think of 'Dr Browne' as a brand name that applied to all his writings. But one of the genres in which *Religio* itself participates—in addition to the essay, meditation, and autobiography—is the familiar epistle, since the work is framed as an answer to a friend's query about the nature of Browne's beliefs. If Browne had taken so seriously one friend's invitation to discuss religion, why not another's? And indeed Browne's ongoing revisions of the *Religio* suggest that his scope has expanded from one special friend to friends that he has not even met.

There is no record of Browne's direct response to Duncon, no extant letter and no mention by Browne of Duncon to anyone else in the extensive correspondence that Browne left. But arguably we do have one indirect response. One further piece of evidence supporting 1659 as the terminus a quo of Duncon's letter is the fact that *Religio Medici* was published twice in that year, once together with the *Hydriotaphia* and *Garden of Cyrus*, but also in its own fifth and annotated edition. Crucially, no further edition of *Religio* would appear until 1669, in which edition one of the passages singled out by Duncon has been seriously altered. Up through 1659, one reads the sentence essentially as it is quoted by Duncon 'that God loves us for that part which is, as it were himself, and the traduction of his holy spirit'. In 1669, this has been changed so that God 'can love nothing but himself, and the traduction of his holy Spirit'.[31] Yet at the same time, Browne retains the three other citations that Duncon has culled out: Moses is still absurd; heaven is not really a specific place; and 'there is a piece of divinity within us, some thinge that was before the elements'. Has the quotation about divine love been altered by mistake? Or was there something a little too bold about claiming that God loves himself in us, or that our

[30] Whitehead, *The Christian progress*, 5–6.

[31] *Religio Medici, The sixth edition, corrected and amended* (London, 1669), 174.

souls are traduced from the Holy Spirit as against from our parents and ultimately from Adam and Eve, the hereditary basis of our fallen state which Browne cannot forget in one of the several discussions of monstrosity in *Religio Medici*? Whatever the case, the publication history of *Religio Medici* confirms that Browne's work had a way not just of appealing to readers of widely ranging religious persuasions, but also of testing, even putting pressure on the boundaries of orthodoxy itself.

A haberdasher by trade, and having grown up in Ipswich, Duncon was politically active against the policies of King Charles from the 1630s, vocal against Royalist abuses in the early 1640s, and an ardent supporter of Parliament in the civil war itself within the county of Suffolk just to the south of Browne's Norfolk.[32] Moving into the 1650s, he was supportive of the Particular Baptists and more generally of the liberty of conscience, an opponent to tithes and a critic of the new government's track record on help to the poor and disabled. Until that postscript in 1659, most of his concerns were practical: how to raise money for soldiers, for instance, or how to keep the peace in local communities. Early in the 1660s, as a young apprentice whose master was 'the Colonel for the City, and one of the chiefest magistrates upon the bench', he was struggling over the question of whether to take the Oath of Allegiance and Supremacy, suffering imprisonment for his refusal but also for opening his shop on a traditional holiday.[33] By 1670, his letters from prison were intensely painful and powerful ones, urging the Norwich magistrates to attend to the human damage and suffering that their persecution was causing the loving, industrious, peaceful and morally upstanding families of their community (Eddington 1932, 72–6). Together with Anthony Alexander, he composes a letter whose plaintive indictments of injustice serve a purpose far different from the letter to Browne: 'Our Opression', he writes,

> is more than we ought always to bear in Silence … And now we are upon the Brink of Ruin … made harborless in our own Houses, and the Widow and the Fatherless have been forced to wander from Place to Place for a Night's Lodging … And what would you have us do? Do you think we are only willful, and resolve so to be?[34]

But Duncon's commitment to Quaker unity and evangelism only increased: in the 1670s, Norwich documents are filled with his activities, attending regular meetings, reaching out in a spirit of unity in the face of adversity to Presbyterians and Independents, his appeals to them based in John Foxe's *Book of Martyrs*, striving to be a good businessman who 'employs several poor people' from the area, facing a mob intent on destroying his shop, helping to open the New Meeting House, of which Duncon was a trustee, trying to convince lapsed Friends to return to the fold, placed in charge of the poor collection, and serving as the chief Norwich distributor of Quaker books from London (Eddington 1932, 80–1, 86–90, 94–8, 133, 138, 226, 252).

---

[32] For the life of Duncon, see Grace, 2004.

[33] See Eddington, 1932, 33, 42, and Joseph Besse, *A collection of the sufferings of the people called Quakers*, vol. 1 (1753), 496 f. In *The second part of the continued cry of the oppressed*, William Penn treated Duncon's persecution in Norwich (87–8). See also Whitehead, *Christian progress*, 484–90.

[34] Besse, *A collection*, 496.

Even after his death, his widow was receiving letters of encouragement from George Fox in the very year, 1682, in which Browne's own death would coincide with the outbreak of one of the worst Norwich bouts of persecution against the Quakers of the whole half-century (Eddington 1932, 182–3).

Given *Religio*'s opposition to persecution, that work's appearance in 1669 might well be taken as criticizing the Restoration oppression of dissent. From that vantage, the small change in the 1669 edition of *Religio Medici* hardly counted as a major rebuke to the Friends. But even if Browne wanted to distance himself from the Quakers, at least one Quaker, and that an especially problematic one, saw fit to feature Browne in print as an exceptional hero of the faith. This Friend was John Perrot. In 1661, in one of his several attempts to convert foreign peoples and powers—the Venetians and the Turks, but also the Pope and Catholic clergy—to the Spirit, Perrot wrote from his prison cell in a Roman madhouse that all 'tender, moderate' Roman Catholics should heed his vision of true religion. Perrot's is an ecumenism toward Catholics that superficially resembles Browne's own in *Religio Medici*. Then, in that general epistle in which he lays out the Quaker principles of the divinity within us, attacking the human forms, lusts, and imaginations that limit the divine: in the middle of this letter, Perrot suddenly notes 'how that in the dayes of Luther, GOD moved in the Darkness upon the face of the deeps thereof; but they which made him their Rest are confounded, and their head is broken as Clay'. Amid this darkness, God 'appeared brighter by Behman and Brown', though this light has been diminished by the fragmentation of truth into sectarian 'confusion'.[35]

The triangulation from Perrot to German mystic Jacob Boehme, then again to Thomas Browne is a messy affair. For one thing, scholars of the Quaker movement have disagreed about whether the German mystic Jacob Boehme was a seminal figure for the Friends; early Quakers themselves were often eager to reject the derivation of their views from Boehme's works, though both in the eyes of critics such as Henry More and in the experience of some Quaker converts such as Anne Conway, the trajectory from Boehme to Quakerism was scarcely unusual.[36] But if Boehme was an uncertain 'fit' in Quaker thought and culture, so too was Perrot himself whose own enthusiasms earned the disfavor of George Fox and eventually got him ousted from the Society of Friends.[37] If Browne was obviously no easy fit into Quakerism either, his *Religio* would never be confused with any of Boehme's works. Perrot may well have had in mind their mutual combination of natural theology—according to which the spirit of God is imminent in the hieroglyphs and signatures

[35] John Perrot, *Battering Rams against Rome* (1661), 11–2. Smith, 1989, 109n, calls attention to this reference in a series of works, making the case that Thomas Browne is the likely candidate, not Robert Browne, the Elizabethan separatist; cf. Smith, 1994, 226.

[36] For the argument against, see Nuttall, 1992, 16. A strong vote in favor comes from Jones, 1914, 208–34. For careful assessments of Boehme's English importance after he was translated beginning in the 1640s, see Smith, 1989, chapter 5; and Cope, 1956, 739–40. For More and Conway, see Hutton, 2004, 65–6. For Perrot, see Smith, 2004; and Carroll, 1970.

[37] See Smith, 1995, 64.

of creation, and external phenomena express eternal realities—with an intense and dynamic spiritualism, according to which God, heaven, hell, paradise and all redemptive means and processes are internal ones, with the two tendencies joined sometimes by alchemy, at other times by a loosely Neoplatonic metaphysics of emanations and effluences from the divine abyss, still other times by Trinitarian patterns or palimpsests.[38] From the vantage point of English intellectual culture, both Dr. Browne and Jacob Boehme had burst upon the scene at the same moment in the 1640s. If Browne made strange bedfellows with English radicals and German mystics, Duncon and Perrot appear to have believed that the Norwich physician could not be contained within the Anglican fold, whilst Browne's other seventeenth-century readers sometimes maintained that he did not fit into the category of Protestant or even of Christian.

Both its theology of Spirit and its outreach to the world meant that Quakerism would accentuate, on the one hand, the essential unity of religious faith and experience and, on the other, their radical individuation. Nor could the status or the basis of unity be a simple matter for the Friends: did unity need translation from language to language, or could the Friends expect the Spirit to translate their English for Native Americans, European Jews, or for that matter the Pope himself? Did unity need language at all or was it better served by silence? If language was crucial, was unity better served by accommodating epistles or by marketplace cries of woe? Did unity need a meeting house? Or was a meeting house just another imaginary form in which the unity of the Spirit was perverted or wholly lost? If the guarantor of unity was elusive for the Quakers, it was no less important—and elusive—in *Religio Medici*, in which the 'sameness' of faith seems variously to depend on Scripture, Church, charity, and reason whilst faith can expatiate in singular dreams and imaginings, best expressed by metaphors, perhaps finding truths never before touched on, yet subject to evaporation in the waking life of Browne himself, threatened by the proximity of heresy, perhaps to be attributed to a ridiculous, melancholy, and ever changing self. In singling out certain passages in Browne's strange *Religio*, Duncon did not so much ignore Browne's testimonials of his loyalty to the Church of England; rather, he took for granted the possibility that in the exchange of letters, a conversation might begin in which Browne would be finally liberated by the Spirit for which he testified from the carnal human forms dressed up as canons and orthodoxies.

References

Barbour, Reid. 1998. *English epicures and stoics: Ancient legacies in early modern culture*. Amherst: University of Massachusetts Press.

Bauman, Richard. 1983. *Let your words be few: Symbolism of speaking and silence among seventeenth-century Quakers*. Cambridge: Cambridge University Press.

[38] For Boehme, see Jones, 1914, chapters 9–12; Walsh 1983; Weeks, 1991; and Koyré, 1929.

Braithwaite, William C. 1955. *The Beginnings of Quakerism*. Revised ed. Henry J. Cadbury. Cambridge: Cambridge University Press.

Brockbank, Elisabeth. 1929. *Richard Hubberthorne of Yealand: Yeoman-Soldier-Quaker, 1628–1662*. London: Friends' Book Center.

Carroll, Kenneth L. 1970. *John Perrot: Early Quaker schismatic*. London: Friends' Historical Society.

Cope, Jackson I. 1956. Seventeenth-century Quaker style. *Publications of the Modern Language Association of America* 71: 725–54.

Corns, Thomas N., and David Loewenstein (eds.). 1995. *The emergence of Quaker writing: Dissenting literature in seventeenth-century England*. London: Frank Cass.

Davies, Adrian. 2000. *The Quakers in English society, 1655–1725*. Oxford: Clarendon.

Eddington, Arthur J. 1932. *The first fifty years of Quakerism in Norwich*. London: Friend Historical Society.

Grace, Frank. 2004. 'Duncon, Samuel (*d.* 1679)'. *Oxford dictionary of national biography*. Oxford: Oxford University Press. Online ed. Jan 2009, http://www.oxforddnb.com/view/article/8244. Accessed 2 May 2011.

Horn, Matthew. 2008. Texted authority: How letters helped unify the Quakers in the long seventeenth century. *The Seventeenth Century* 23: 290–314.

Hutton, Sarah. 2004. *Anne Conway: A woman philosopher*. Cambridge: Cambridge University Press.

Ingle, H.Larry. 1994. *First among friends: George Fox and the creation of Quakerism*. Oxford/New York: Oxford University Press.

Jones, Rufus M. 1914. *Spiritual reformers in the 16th & 17th centuries*. London: Macmillan.

Koyré, Alexandre. 1929. *La Philosophie de Jacob Boehme*. Paris: Vrin.

McDowell, Nicholas. 2003. *The English radical imagination: Culture, religion, and revolution, 1630–1660*. Oxford: Clarendon.

Moore, Rosemary. 2000. *The light in their consciences: Early Quakers in Britain, 1646–1666*. University Park: Pennsylvania State University Press.

Nuttall, Geoffrey F. 1992. *The holy spirit in Puritan faith and experience*. Chicago: University of Chicago Press.

Peters, Kate. 1995. Patterns of Quaker authorship, 1652–56. In *The emergence of Quaker writing: Dissenting literature in seventeenth-century England*, ed. T.N. Corns and D. Loewenstein, 6–24. London: Frank Cass.

Platt, Joan. 1926. *The Quakers in Norwich*. Norwich: Goose & Son.

Reay, Barry. 1984. Quakerism and society. In *Radical religion in the English revolution*, ed. J.F. McGregor and B. Reay, 141–64. Oxford: Oxford University Press.

Reay, Barry. 1985. *The Quakers and the English revolution*. New York: St. Martin's.

Smith, Nigel. 1994. *Literature and revolution in England, 1640–1660*. New Haven /London: Yale University Press.

Smith, Nigel. 1995. Hidden things brought to light: Enthusiasm and Quaker discourse. In *The emergence of Quaker writing: Dissenting literature in seventeenth-century England*, ed. T.N. Corns and D. Loewenstein, 57–69. London: Frank Cass.

Smith, Nigel. 1989. *Perfection proclaimed: Language and literature in English radical religion, 1640–1660*. Oxford: Clarendon.

Smith, Nigel. 2004. 'Perrot, John (*d.* 1665)'. *Oxford dictionary of national biography*. Oxford: Oxford University Press. Online ed. May 2007, http://www.oxforddnb.com/view/article/21987. Accessed 6 May 2011.

Vann, Richard T. 1969. *The social development of English Quakerism, 1655–1755*. Cambridge, MA: Harvard University Press.

Walsh, David. 1983. *The mysticism of innerworldly fulfillment: A study of Jacob Boehme*. Gainesville: University Press of Florida.

Weeks, Andrew. 1991. *Boehme: An intellectual biography of the seventeenth-century philosopher and mystic*. Albany: State University of New York Press.

Chapter 4
Writing Authority in the Interregnum: The Pastoral Letters of Richard Baxter

Alison Searle

4.1 Introduction

Approximately 1,300 letters survive in the archive of the correspondence of Richard Baxter (1615–1691). A large percentage of these date to the 1650s and incorporate individuals from all levels of society: London apprentices, ministers, scholars, the gentry and aristocracy. This essay explores the construction of pastoral authority within the fluid and dynamic social context of the Interregnum through four brief case-studies: Baxter's correspondence with an imprisoned Scottish nobleman, the Earl of Lauderdale, a Derbyshire gentlewoman, Katherine Gell, a Cambridge student, Thomas Doolittle and a young chaplain, Abraham Pinchbecke. The vacuum created by the collapse of traditional religious and political structures, as well as the spirit of experimentation that the Interregnum fostered, shapes the epistolary discourse of each of these individuals in important ways. The disintegration of established hierarchies of power enabled the development of Baxter's unique pastoral authority; this was achieved, in part, through the medium of the letter.[1] Baxter's correspondence with these four individuals reveals the significance of social status and gender as they inflect pastoral epistolary discourse; the importance of diverse communities (centred around the local parish, print publication or even political imprisonment)

I am grateful to Dr Jean Lambert for reading and commenting on an earlier draft of this essay.

[1] John Spurr, *The Post-Reformation: Religion Politics and Society in Britain 1603–1714* (Harlow: Pearson Longman, 2006), 133–4.

A. Searle (✉)
Department of English, University of Sydney, NSW Australia, 2006
e-mail: alison.searle@sydney.edu.au

A. Dunan-Page and C. Prunier (eds.), *Debating the Faith: Religion and Letter Writing in Great Britain, 1550-1800*, International Archives of the History of Ideas 209,
DOI 10.1007/978-94-007-5216-0_4,

to the formation and maintenance of letter-writing networks,[2] and the perennial tension between written presence and physical absence that so often recurs in discussions about letter-writing as performance.

James How has argued that it was only after the death of King Charles I and the improvements in communication resulting from the Civil Wars that a public Post Office became a viable reality in England; the Cromwellian regime valued a centralized postal network as it facilitated effective political surveillance.[3] How suggests that the Post Office created 'A space of mind no longer containing 'striving souls' and all the paraphernalia of the spiritual world but instead the dazzling shops and theatres and brothels of London; the dressing rooms of intimate but distant friends; and potentially rewarding access to the inner offices of the great and powerful'. This new epistolary space was 'a different kind of venue for being; or being in action' focused on 'the physical and not the spiritual world' (How 2003, 98–99). In a crucial sense, however, Baxter's network of correspondents demonstrates the establishment and exploration of precisely the kind of epistolary space that How argues was 'collapsing' during this period. To some extent his account endorses the political and ecclesiastical settlement enforced by the Restoration which pushed the epistolary discourse of Baxter and many of his correspondents into the problematic space defined by nonconformity or dissent, maintained by espionage, surveillance, parliamentary hostility and legal sanctions after 1660.[4] This analysis of the interplay between social, political and religious upheaval, Baxter's writing practice and his construction of authority (or a public persona capable of exercising significant power in a sense that can best be defined as pastoral) through print publication and epistolary networks in the 1650s involves a partial reversal of the secularisation narrative that How tells.

Baxter's epistolary activities during the Interregnum can be read, in part, as driven by and resulting from, his experiences during the Civil War and his attempt to forge an interim political, ecclesiastical and pastoral solution to the chaos engendered (in his diagnosis) by a combination of violence and the heady

[2] Neil Keeble has noted how crucial letter-writing was to the formation and maintenance of 'a sense of community and fellowship despite being forcibly separated' amongst nonconformists following the Restoration. As Baxter's correspondence networks make clear, however, this was also significant during the unsettled and inchoate period of the Commonwealth and Protectorate characterized as it was by constant shifts in the political and ecclesiastical settlement of Church and state. *The Literary Culture of Nonconformity in Later Seventeenth-Century England* (Leicester: Leicester University Press, 1987), 78–82.

[3] James How, *Epistolary Spaces: English Letter Writing from the Foundation of the Post Office to Richardson's Clarissa* (Aldershot: Ashgate, 2003), 4–12. For continuities between the Commonwealth and the restored Stuart Regime and the central role that the Post Office played in covert surveillance see Alan Marshall, *Intelligence and Espionage in the Reign of Charles II, 1660–1685* (Cambridge: Cambridge University Press, 1994), 4–6, 18–27, 78–95.

[4] Marshall, 8–10, 78–95.

doctrine of 'justification by free grace alone through faith alone' (Cooper 2001, 197).[5] This biographical trajectory is one possible reconstruction of Baxter's activities during the 1650s linking his initial forays into print, specifically *Aphorisms of Justification* (1649) and *The Saints' Everlasting Rest* (1650); his ceaseless advocacy of ecumenism and Church unity—finding its strongest practical expression in the establishment of ministerial associations in Worcestershire and other counties throughout England that embraced a spectrum of Protestant religious opinion[6]; his increasing political activism, particularly in the period immediately prior to the Restoration; and his extraordinarily prolific epistolary output. Letters could be the catalyst for action, the means of enactment, or the unexpected result of Baxter's intervention in the public sphere. To some extent they enable us to trace the ways in which fear and idealism helped to construct a concept of pastoral authority, enacted through Baxter's letters, as much as in his parochial ministry at Kidderminster,[7] in a period of unique, but limited opportunities from 1649 to 1660.

Baxter's authority on the national stage was initially established through the writings he published during the Interregnum. His personal, conversational and direct style of discourse implicitly courted a wide variety of epistolary responses: these could be admiring, contradictory, outraged, interrogatory, grateful, troubled, or, at times, a complex combination of any of these. Baxter replied to all his correspondents, as efficiently and thoroughly as he was able: each time his pastoral authority—as a controversialist, professional mentor, colleague or casuist—was more firmly inscribed and broadly disseminated throughout his epistolary network of scholars, clergymen, godly gentry, London tradesmen and local associations of Churches. Baxter's successful parochial ministry in Kidderminster during the 1650s, particularly his practice of catechising families individually, publicized in his widely-read *Gildas Salvianus: The Reformed Pastor* (1656), cemented his reputation as a pastoral authority and encouraged further correspondents from all sectors of society to write to him. It was this writing of authority during the Interregnum—public, pastoral and epistolary—that helped to create Baxter's status as a leader in the Puritan movement prior to the Restoration and which underwrote his ongoing influence as a nonconformist after the Act of Uniformity in 1662.

[5] William Lamont, *Richard Baxter and the Millennium* (London: Croom Helm, 1979), 124–209; Tim Cooper, *Fear and Polemic in Seventeenth-Century England: Richard Baxter and Antinomianism* (Aldershot: Ashgate, 2001); Tim Cooper, 'Why Did Richard Baxter and John Owen Diverge? The Impact of the First Civil War', *Journal of Ecclesiastical History* 61.3 (2010): 496–516.

[6] A full discussion of Baxter's ecclesiology is provided by Paul Lim, *In Pursuit of Purity, Unity and Liberty: Richard Baxter's Puritan Ecclesiology in its Seventeenth-Century Context* (Leiden: Brill, 2004).

[7] For connections between Baxter's pastoral ministry in the 1650s and his compendious *Christian Directory* see John Brouwer, 'Richard Baxter's *Christian Directory*: Context and Content' (unpublished PhD thesis, University of Cambridge, 2005).

4.2 The Earl of Lauderdale

The fact that an extensive and intimate correspondence existed between the Earl (later Duke) of Lauderdale, John Maitland (1616–1682) and Baxter, is rather surprising.[8] However, as was so often the case with Baxter's correspondents, the Earl was prompted to write from Windsor Castle on 19 October 1657, by an appreciation for Baxter's theological writings.

> Give leave, Sir, therefor in the first place to acknowledg myself your debtor for my share of your charity to mankinde—I meane for the good which by the grace of God I have gott by your most pious and learned Labours: especially your book of Rest which I have reason to esteem above all bookes except the Bible.[9]

The precipitating factor for their epistolary exchange was thus Baxter's early foray into religious print publication. Only Lauderdale's side of the correspondence has survived—this section focuses on the fourteen letters he sent Baxter between 19 October 1657 and 31 March 1660. For most of this period Lauderdale was imprisoned at Windsor Castle. His enforced leisure prompted him to offer Baxter his assistance in translating any works out of French, Italian or Spanish, that he might require for his future writing projects. Baxter took him up on this offer and publicly acknowledged his indebtedness in his preface to *A Key for Catholics* (1659): '*Though I understand not the* French *tongue*' by '*the help of a Noble friend that hath vouchsafed to translate some part of them for my use, I am imboldened to a confidence*' that Cardinal Jacques Davy Duperron has been effectively refuted by David Blondel and Petrus Molinaeus (Keeble and Nuttall 1991, vol. 1, 355). Lauderdale offers to obtain books for Baxter in Holland or Paris; he attempts to locate and translate relevant sections of scholarly works, describing his explorations (whilst imprisoned) of various libraries at Eton, and deputising a friend to try and locate a book at St Paul's, in London; he recommends Baxter's works to a friend overseas, reporting his opinion back to Baxter[10]; he urges Baxter to arrange for the translation of his own books, especially *The Saints' Everlasting Rest* into Latin, so that it can become more widely available; and at times, he cautiously attempts to engage Baxter's assistance, both in relation to his own imprisonment, the sequestration of his property and the restoration of Charles II.

---

[8] See Frederick J. Powicke, 'Eleven Letters of John Second Earl of Lauderdale (and First Duke), 1616–1682, to the Rev. Richard Baxter (1615–1691), *Bulletin of the John Rylands Library* 7 (1922–1923): 73–105 for further discussion of this correspondence.

[9] Dr Williams's Library (DWL), MS 59.IV.104. This refers to the manuscript collection of Richard Baxter's correspondence held at Dr Williams's Library and I am grateful to the Trustees for permission to quote from them here. In transcribing the letters I have reproduced the capitalisation and spelling in the original; common abbreviations have been silently expanded; all other expansions, or uncertain readings have been enclosed in square brackets.

[10] This was probably Alexander Lindsay, 1st Earl of Balcarres, and Lauderdale's cousin. N. H. Keeble and G. K. Nuttall, *Calendar of the Correspondence of Richard Baxter*, vol. 1, 362. Baxter later had a close and profitable acquaintance with his wife.

The regular correspondence, though not the relationship, between Lauderdale and Baxter petered out on 31 March 1660. It is possible that Lauderdale's physical imprisonment by the various Interregnum regimes created a space of enforced impoverishment and idleness that caused him to turn uncharacteristically to scholarly pursuits, the cultivation of a deeper spirituality, and the consolation of epistolary discourse and friendship with one who had authored a number of books that gave him comfort during his imprisonment.[11] It may be that the loss of his political, social, religious and military status (characteristic of Lauderdale's position both prior to 1651 and after 1660), accompanied by his imprisonment, resulted in the deconstruction of his powerful masculine persona—his national, public identity. The concept of gender identities only becomes problematic in this period when something happens to throw them into question. Partly by unavoidable circumstances, partly by choice, during the period of his correspondence with Baxter, Lauderdale was disempowered, confined to a kind of 'domestic' space of limited movement, social intercourse, devotion and mediated communication, more usually associated with women. The unravelling of traditional political and religious structures, including the Established Church and a powerful royalist nobility, alongside the proliferation of religious print publishing during the Interregnum, thus coalesced to forge a unique epistolary encounter and friendship in the lives of Baxter and Lauderdale: to some extent the 'feminisation' of the latter created the pastoral authority of the former; both roles were constructed, defined and maintained solely through the exchange of letters.

4.3 Katherine Gell

Like Lauderdale, the Derbyshire gentlewoman, Katherine Gell (1624–1671), initiated a correspondence with Baxter after she had read *The Saints' Everlasting Rest*. She begins, in July 1655, by observing: 'Though I never yet saw \you/ I hope having seene soe much of you it may be an incouragement to me to write for some satisfaction from \you/ in somthing of your owne writing in your Saints rest'.[12] Gell believes that her acquaintance with Baxter's spiritual writings in print has facilitated a large measure of acquaintance—'having seene soe much of you'—and she seeks to extend this familiarity through the mode of epistolary discourse. Unlike Lauderdale, however, Gell does not simply wish to thank Baxter, nor offer

[11] Gary Schneider has commented on 'the crucial function of letters in communicating with prisoners. That letters overcame physical barriers was not an imaginative construct, but a literal fact'. He quotes Robert Southwell who asserts: 'It hath bin alwaies a laudable custome in Gods Church, for such as were afflicted in time of persecution … by letters and bookes, to comfort one an other', *The Culture of Epistolarity: Vernacular Letters and Letter Writing in Early Modern England, 1500–1700* (Newark: University of Delaware Press, 2005), 188–9.

[12] MS 59.V.216.

him the scholarly services of book provision and translation. Her sensitive conscience has been disturbed by his injunction to 'earnestly presse the helpeing of others to your rest & in particular by admonishing I was soe clearly convinced of my neglect & unaptnes to it & my uselessnes in my family that way; that I have drawne this conclusion from it that I am not in a state of grace'.[13] Her natural bashfulness renders her incapable of speaking easily and freely with others regarding religious issues—the medium of the letter, with its implicit valuation of absence over presence is thus a positive factor. She turns to Baxter as a spiritual pastor and experienced casuist, pleading that he will 'excuse both stile English & all other defects herein by considering its a womans'.[14] Gell's disparaging reference to her gender can be read as a concession to the prevailing stereotype of female weakness held by many of her contemporaries, including Baxter, who observes in a later letter: 'most women are of more sensible passionate dispositions than men'.[15] However, paradoxically, her concept of gender is also enabling, as it allows her freely and unapologetically to address Baxter by letter over the concerns which his spiritual advice, publicly disseminated through print, has raised for her personally; Baxter's writing style, even in print, was such that it implicitly invited his readers to respond to him. It is possible that Gell's willingness to assume in part her period's stereotype of female weakness allowed a greater degree of freedom in epistolary discussion of her spiritual concerns than Lauderdale demonstrates even when most disempowered and spiritually susceptible.[16] Social status, as well as gender, may have shaped the degree to which Lauderdale was willing to seek pastoral counsel. Gell was the daughter of a prominent Puritan administrator, John Packer, and married into a family of equally strong godly convictions—the Gells of Hopton, Derbyshire.[17] Her letters, however, record the disgust of some of her family and acquaintance amongst the gentry at the thought of communing with a clergyman

[13] MS 59.V.216.

[14] MS 59.V.216.

[15] MS 59.V.11.

[16] It is important to note that early modern and contemporary scholarly assumptions about the constraints placed upon women at the time and their relegation, by and large, to the 'private' sphere fail to acknowledge the breadth and intelligence with which women, particularly Puritan women, engaged in the intellectual culture of the period. See especially Johanna Harris and Elizabeth Scott-Bauman, 'Introduction', *The Intellectual Culture of Puritan Women, 1558–1680* (London: Palgrave Macmillan, 2010), 1–9. Gell also participated in these networks, but it is beyond the range of the current discussion to explore this aspect of her writings. See Keith Condie, 'The Theory, Practice, and Reception of Meditation in the Thought of Richard Baxter' (unpublished PhD thesis, University of Sydney, 2010), 323–58; Alison Searle, ''My Souls Anatomiste': Richard Baxter, Katherine Gell and Letters of the Heart' in *Early Modern Literary Studies* 12.2 (2006) for further details of this correspondence.

[17] William Lamont, 'Gell, Katherine (*bap.* 1624, *d.* 1671)', *Oxford Dictionary of National Biography*, Oxford University Press, 2004 [http://0-www.oxforddnb.com.catalogue.ulrls.lon.ac.uk/view/article/66997, accessed 15 March 2011]

lower than a bishop.[18] Thomas Doolittle and Abraham Pinchbecke, who were both of middling social status and in positions of dependency (as a student and household chaplain respectively) had little reservation in engaging Baxter in an epistolary discussion about the condition of their souls.

Baxter's correspondence with Gell lasted until 27 December 1658. It is possible to reconstruct the relationship in some detail, as both sides of the exchange have survived.[19] Gell is, in general, deferential and anxious, relying on Baxter both for pastoral reassurance as to her state of grace and for counsel in a range of matters including usury, fear of the dark, religious affections, prayer and the eternal destiny of young children (one of hers died at seventeen weeks of age). Baxter's epistolary response is timely, personal, occasionally patronising, but in general appropriate. The letter, as a genre, was integral to his pastoral practice and care. His letters to Gell are very similar to those he wrote to his future wife, Margaret Charlton, when she was his parishioner at Kidderminster and lived only a short walk from him. His letter-writing also had an important influence on the literary style and spiritual counsel of his print publications. For Baxter, the medium was of little significance when it came to exercising pastoral authority and providing godly counsel. The consistency of his literary style and advice and the ways in which his epistolary practice shaped his authorial persona can be demonstrated by comparing his advice to Gell regarding her spiritual dullness and depression (7 June 1656), with his biographical account of his wife, Margaret Baxter, published some 35 years later in 1681. Baxter writes to Gell:

> Sit not at home as if you had no body to looke after but yourselfe; but step out now and then to your poore tenants, or send for them to you, and deale with them about the matters of their salvation. And you may find that, compassion to them, and such holy discourse in the worke of God, will do more to enliven you, then much Sorrow Striving with your heart will do.[20]

He observes of his wife:

> When we were married, her sadness and melancholy vanished; counsel did something to it, and contentment something; and being taken up with our household affairs, did somewhat. And we lived in inviolated love, and mutual complacency, sensible of the benefit of mutual help.[21]

The affinities with his religious publications, addressed to a far wider audience, can be seen in the same letter to Gell. Baxter refers her to his book *The Right Method for a Settled Peace of Conscience* (1653), which had itself been written in response to the spiritual anxieties of one of his female parishioners at Kidderminster,

---

[18] MS 59.V.5.

[19] Keith Condie, 'Some Further Correspondence between Richard Baxter and Katherine Gell' *The Historical Journal* 53.1 (2010): 165–76.

[20] MS 59.V.218.

[21] Richard Baxter, *A Breviate of the Life of Margaret, The Daughter of Francis Charlton ... Wife of Richard Baxter* (1681), 47.

Margaret Bridges.[22] His letter reassures Gell in language remarkably similar to that used in print for his female parishioner and a wider, unknown public audience: 'Though lively affections and Sensibility be very desirable yet are they not the evidences by which the truth of Grace may so well be tried'.[23] During the Interregnum, with the disestablishment of the state Church and the flourishing of religious publications in print, Baxter's pastoral methods and counsel (so effectively implemented in Kidderminster) reached a far wider public audience through his published writings. The accessibility and warmth of his authorial persona encouraged an active response on the part of his readers (such as Lauderdale and Gell); consequently the reach of his parish and pastoral authority extended beyond Kidderminster to Derbyshire and Windsor Castle through the medium of the letter. The absence of a state Church strictly defined by geographical location and ecclesiastical hierarchy allowed (perhaps necessitated) experimental and imaginative conceptions of Christian fellowship and pastoral care; these could be and were enacted, to some extent, through the medium of the letter.

4.4 Thomas Doolittle

Baxter also had an intense concern for the education and training of future ministers. This found expression in his repeated exhortations to the rich to contribute towards supporting students[24]; his endorsement of Matthew Poole's innovative, but eventually abortive, scheme to provide permanent maintenance for ministerial trainees[25]; and his own active pursuit of appropriate candidates, provision for them at university, and practical training of young assistants before they were assigned parishes of their own.[26] Baxter's specific concern for the training of ministers was a reflection of his broader passion for the education and development of youth.[27] One key expression of this interest is Baxter's epistolary nurture of young men, like Thomas Doolittle (1630/1633–1707), when preparing for, or establishing

[22] Richard Baxter, 'To the Poor in Spirit,' *The Right Method for a Settled Peace of Conscience* (1653).

[23] MS 59.V.217.

[24] For example, the dedicatory epistle to Thomas Foley and 'The Preface: To the Nobility and Gentry, and all that have the Riches of this world' in Richard Baxter, *The Crucifying of the World by the Cross of Christ* (1658). See also MS 59.I.127, IV.214.

[25] Richard Baxter, 'To the Rich that love Christ, the Church, the Gospel, and themselves' in Matthew Poole, *A Model for the Maintaining of Students of Choice Abilities at the University* (1658).

[26] Richard Baxter, *Reliquiae Baxterianae*, I, 89; J. William Black, *Reformation Pastors: Richard Baxter and the Ideal of the Reformed Pastor* (Milton Keynes: Paternoster, 2004), 238–41.

[27] See for example MS 59.III.257, IV.65, IV.91.

themselves in, their callings.[28] This became particularly imperative during the 1650s when the existence of a state Church of any kind was an issue of rancorous public debate.

Baxter deliberately models his epistolary relationship with Doolittle on the example provided by the apostle Paul to Timothy. In 1678 he wrote a prefatory letter to Doolittle's book, *The Protestant's Answer*, where he describes him as: '*My Son and fellow-servant in the work and patience of the Gospel*'. This echoes Paul's address to Timothy: 'my own son in the faith' (1 Timothy 1:2); 'my dearly beloved son' (2 Timothy 1:2); and his description of him as a co-labourer when writing to the Church in Philippi, 'Paul and Timotheus, the servants of Jesus Christ' (Philippians 1:1).[29] The Doolittle family were prominent in Kidderminster; Thomas had been converted under Baxter's ministry, when he was preaching the series of sermons that formed the foundation of his devotional classic, *The Saints' Everlasting Rest*. Doolittle was initially employed by a local lawyer, but when he was required to work on a Sunday he rejected the law and decided to pursue pastoral ministry.[30] Baxter thus knew his family, was instrumental in his conversion and, as the correspondence exchanged between them clearly demonstrates, supported his studies towards a BA and then MA at Pembroke Hall, Cambridge.[31] When Doolittle graduated Baxter found a position for him as a chaplain; however, he had already been invited to become pastor at St Alphage, London.[32] Baxter continued to be an important mentor to Doolittle in the early stages of his ministry, sending him books, recommending reading material and providing theological advice on contemporary issues of acrimonious controversy.[33] The relationship eventually shifted to one of professional solidarity and respect, as is evidenced by the fact that Baxter publicly recommended Doolittle's writing and privately encouraged several distressed individuals who wrote to him to seek Doolittle's pastoral advice.[34]

Baxter's epistolary nurture of Doolittle played an important role in what J. William Black has described as 'the mentoring relationships that often supplemented a 'godly' young man's preparation for ministry' (Black 2004, 241).[35]

---

[28] Black examines Baxter's correspondence with Doolittle at length, but he focuses on the way in which Baxter mentored Doolittle as a student and young pastor, rather than on the specific role that letters as a genre contributed to this development, 236–54.

[29] Black, 246–7. For a discussion of the importance of the biblical epistles to the practice of letter-writing more generally in the early modern period see C. A. Patrides, 'The Epistolary Art of the Renaissance: The Biblical Premises', *Philological Quarterly* 60.3 (1981): 357–67.

[30] Black, 242–3.

[31] MS 59.VI. 28.

[32] Black, 243.

[33] MS 59.VI.28, I.121, I.125, I.123.

[34] Richard Baxter, 'Epistle to the Reader,' *The Protestants Answer to that Question, Where Was Your Church Before Luther?* (1678); MS 59.IV.264, IV.231.

[35] See also Tom Webster, *Godly Clergy in Early Stuart England: The Caroline Puritan Movement, c.1620–1643* (Cambridge: Cambridge University Press, 1997).

Doolittle himself observed that Baxter 'saw fit to send letters distinguished by marks of love and filled with advice about the method of study and to supply help in many ways'.[36] Writing to Baxter from Cambridge on 12 November 1652, Doolittle asks for his counsel: he is about to complete his undergraduate degree and wants to do an MA too. He reveals to Baxter the tension he feels between a longing 'to bee doeing for god & to bee instrumentall in his hands for the good of soules' and his sense that 'the worke is great & my strength but weakenes'; further study will enable him to better prepare for the 'weighty calling which I have in my eye'.[37] He describes his heart as 'farre more hard then many bones I have read off which if they had beene steeped in vinegar & ashes would have lost their nature & become soft that they may be cut with a thread'. Doolittle has attempted to apply the standard pastoral remedies to his own soul without effect and this drives him, like Gell, to Baxter, the experienced casuist: 'oh that I had your answeare here now! which I hope you will Hasten up[on]. And my prayers to god shall bee that he would breath in you by his spirit & dictate & suggest unto your heart, that you might speake home & plaine to my very soule'.[38] Here again, Baxter is being called upon to perform the role of parish preacher and pastor, through the medium of the letter, in an Interregnum context, when both the parish system and the career opportunities of university-trained ministerial candidates had been radically overhauled. Doolittle is deeply appreciative of what Baxter has done for him and to this end not only confesses his personal struggles, but also shares the blessing of God upon his work. The narrative encapsulated in his letter is specifically designed to encourage Baxter to continue the work of seeking out and mentoring young men in order to 'obtaine the ends' he aims at: the glory of God, the extension of his kingdom and the saving of souls.[39] Doolittle himself shared this concern and set up an Academy, following the Restoration, in order to train young nonconformists, who were no longer able to graduate from Oxford or Cambridge.[40]

The remaining three letters date from the earlier part of 1657; they centre upon the issue of 'Universall Redemption'. Baxter's letter of 6 March is a response to some questions in a letter (not extant) from Doolittle, which argues in significant detail the issues of the extent of Christ's death and his intercession.[41] Doolittle replied on 9 May—he has again received several books from Baxter and is grateful

[36] Thomas Doolittle, *The Lord's Last-Sufferings* (1681), translated from the Latin by Black, 247.

[37] MS 59.VI.128.

[38] The kind of support and counsel Baxter offered to university students during their preparation for the ministry can be seen in his correspondence with William Duncumbe of King's College, Cambridge. Duncumbe was particularly concerned about the spiritual and practical care of his sister and seeks Baxter's assistance with this. See MS 59.VI.153, V.226, IV.212. V.189, VI.138.

[39] MS 59.VI.28.

[40] J. W. Ashley Smith, *The Birth of Modern Education: The Contribution of the Dissenting Academies* (London: Independent Press, 1954), 24–5, 284–6; Irene Parker, *Dissenting Academies in England* (New York: Octagon Books, 1969), 138.

[41] MS 59.I.121.

'for your paines concerning the Question mentioned in my Last'. It is not merely of academic interest. The current 'controversy of the universality of Christ Redemption & the extent of his death' has direct implications for the way in which he presents the gospel to his congregation. For if Christ has not, in some sense, died for all, it is probable that in a 'greate Large full congregation there might be one Reprobate … now How can I preach this'—'if thou beleeve thou shalt be saved'—'& assure them in the name of god … if in noe sense Christ dyed for them'.[42] While Doolittle has a real passion for the 'discovery of any truth' and an interest in theological controversy, a right understanding of the matter he raised in his letter to Baxter will enable him to preach more evangelistically and effectively. Though deferent, he has not been entirely persuaded by Baxter's argument and raises several points desiring further clarification. This marks a shift in the power dynamics of the epistolary exchange and the tutelary hierarchy: Doolittle is beginning to develop into a mature fellow-pastor, who is willing not only to seek advice, but also to challenge arguments. In an enlightening aside he quotes Baxter to himself, revealing his own humility and Baxter's openness to dialogue: both necessary preconditions for the kind of epistolary relationship they shared.

> Sir if it doth not become such a pygmy as I to reply to you I bes[eech] you excuse me, for it is out of a desir that I have to be informed in [the] truth, for you blame young students for taking things soe much upon trust from the bare aut[h]ority of others.

And that includes Baxter. Doolittle still looks to him as a mentor: he desires his prayers and confesses his continuing struggles with pride and despondency. Again he shares the success of his ministry which is 'causing my doctrine to fasten upon the hearts of many'. He tells Baxter this 'that god may have the praise from you as well as from mee, & for the refreshing of your soule, whose end was this in doeing what you did for mee'. In a postscript Doolittle reminds Baxter of a 'Catalogue' he has compiled, mentioned in an earlier letter, that listed 'good authors'. He requests Baxter to send him a copy: 'because I intend to furnish my selfe more, & I would that those that I buy should be choise'.[43] Theological argument with the intent of sharpening pastoral ministry; spiritual counsel on matters of heart sin; encouragement in his early years as a preacher; recommendations for reading; and, above all, an honest, discreet and faithful friend, are the elements of epistolary nurture that Doolittle sought from Baxter.[44] These letters played a crucial role in his formation as a student, pastor and teacher; they also demonstrate the construction of Baxter's authority and influence through letters as a pastoral role model, mentor and patron during the Interregnum.

---

[42] MS 59.I.125.

[43] MS 59.I.125–6. Baxter later provided such a list in *A Christian Directory* (1673). Keeble and Nuttall, *Calendar*, vol. 1, 251.

[44] MS 59.I.125.

4.5 Abraham Pinchbecke

Such epistolary relationships, constructed around a concept of pastoral authority and spiritual nurture and facilitated through the medium of the letter were not necessarily rooted in a shared parish context (such as Kidderminster), or a history of financial support through education (as Doolittle had experienced at Cambridge). The young chaplain, Abraham Pinchbecke (1626–1681/1682), like many of Baxter's other correspondents, initiated an epistolary exchange on the basis of a familiarity with his printed works—in this case Baxter's two earliest publications *Aphorisms of Justification* and, more famously, *The Saints' Everlasting Rest.* He worked in the household of the Earl of Musgrave and on 24 June 1653 sent a letter to Baxter constructing an epistolary relationship between them modelled on a divine school: 'I have first begd of God to be my teacher, & he as my master hath sent me to you (if I be not much mistaken) as to an higher scholler in the same schoole with mee to tell mee what I asked'. He then requests Baxter for arguments 'to prove the divinity of the scriptures'. Pinchbecke thinks that this will be the most efficient way to resolve his uncertainties, as he is too sickly to research and read about the topic. He concludes: 'Sir if now you have read my letter you thinke it too troublesome or bold take it & burne it but first pray for its author'.[45]

Baxter's response is characteristically encouraging: 'The Members of Christ have such Interest in each other, & the sight of their faces makes so small an addition, that you needed no Apologie for writing to a stranger'. For Baxter, epistolary discourse is simply an extension, or alternative form, of Christian fellowship and communion. It does, however, pose limitations. He is uncertain whether Pinchbecke needs arguments only for preaching, or whether he himself is 'assaulted with any Temptations to doubt' the divinity of scripture. In order to cover both possibilities, Baxter outlines several arguments at great length. He justifies this on the grounds that if he writes as briefly as Pinchbecke requires 'the arguments will not be seen in their full force & so will be lost through a defective reception'.[46] He encourages Pinchbecke to read several works he cites on the topic, but concludes on a pastoral note: 'If God sent you to mee for Arguments, I hope you may received these as returned from him: Only pray for the eye salve of the Spirit. And againe let one Good Argument seem sufficient'. The difficult work is in 'answering … exceptions' and Baxter suggests that 'one at hand' will be able to do this more effectively than he can.[47]

The constant changes in government characteristic of the Interregnum, and the consequent political and ecclesiastical uncertainty that resulted, exercised Pinchbecke and find expression in his response to Baxter's argument that the perfection of the Mosaic Law is an evidence for the divinity of scripture. He writes:

> doe not we see in Commonweales how after ages perceiving the faileing of theire people in the observation of some precedent laws doe ad remedies to such transgressions & new conditions to old laws, [we] make new laws where the other are imperfect, or the present

[45] MS 59.IV.166.

[46] MS 59.IV.51.

[47] MS 59.IV.52.

condition of affaires requires. & for invention have we not some whose braines have brought forth some new formes of government very rationall?

An exploration of scriptural apologetics can easily become a discussion of 'new formes of government very rationall'.[48] But this was not Pinchbecke's main end in writing to Baxter: he hoped to be quickened to 'greater diligence & zeale in the worke of God'. Though their acquaintance is 'but in a literall discourse' he anticipates that Baxter will be God's instrument to enable him to be more humble and self-denying. He instructs Baxter to 'deale roundly with mee & use as sharpe a style as you please so it be christian & syncere'. Like Doolittle, openness, accountability and honest dealing are the qualities he appreciates and actively looks for in his correspondence with Baxter.[49]

In his reply, Baxter again deals in detail with the objections that Pinchbecke raised to his arguments concerning the divinity of scripture, contending with his 'talke \what/ following ages may do by way of inventing supplementall lawes which is not that I speake of'.[50] His readiness to respond so rapidly and comprehensively to theological issues raised by a young man he does not know reflects how serious an 'Interest' he took in 'the Members of Christ'.[51] Baxter also expects of others the high standards he set for himself. He encourages Pinchbecke to locate 'Grotius, Camero, or Mornay' and read them—'they are common bookes as most in the shops'. The concern Pinchbecke raises about length is given short shrift: 'I say you must not thinke to have any considerable knowledge in any Art or Science, without labour'. Again, later, when outlining the argument for the divinity of scripture on the basis of its wide acceptance amongst early Churches throughout the world, Baxter states: 'Your ignorance of Antiquity must be cured by diligent reading & faithfull endeavour to know what you know not. how you can thinke that I should shew you in any other way, or in a few wordes … I cannot imagine'. This commitment to diligent study grew out of Baxter's conviction of the importance of education in order to equip one to pursue their calling; indeed, he saw it as a privilege of the ministerial office.[52] He states this quite bluntly to Pinchbecke: 'No knowledge

[48] See Lamont, 173–5, for this aspect of the correspondence. He argues that Baxter responded increasingly to the excitement and opportunities offered by the Commonwealth and Protectorate, in part influenced by his correspondence with men such as Henry Oasland and John Dury. The ultimate outcome, Lamont asserts, was Baxter's *Holy Commonwealth* (1659) dedicated to his ideal magistrate, Richard Cromwell.

[49] MS 59.IV.43.

[50] MS 59.IV.170.

[51] MS 59.IV.51.

[52] 'Its something that you are maintained by other mens labours, and live on the common-wealths allowance…. Is it nothing to be bred up to Learning, when others are bred at the plough and cart? and to be furnished with so much delightful knowledge, when the world lieth in ignorance?' Richard Baxter, *Gildas Salvianus, The Reformed Pastor* (1656), 253–54. Pinchbecke's professed ignorance does not result from any educational gap in comparison with Doolittle. He attended St Catharine's College, Cambridge, gaining his BA in 1647 and his MA in 1651. William Lamont, 'Pinchbecke, Abraham (*bap.* 1626, *d.* 1681/2)', *Oxford Dictionary of National Biography*, Oxford University Press, 2004 [http://0-www.oxforddnb.com.catalogue.ulrls.lon.ac.uk/view/article/67004, accessed 14 March 2011]

will be got without labour you are made for nothing els … but to know Gods will & do it'. Even though he is generous in the time he expends answering Pinchbecke's queries, Baxter refuses to allow his summaries and explications to be seen as a substitute for careful and thorough study of the sources by Pinchbecke himself.[53]

Pinchbecke is appreciative of the good that Baxter's letters 'both as argumentative & exhortatory' have done for him. He requests the freedom to continue to address his 'scruples & irresolutions doctrinall & practicall' in correspondence with Baxter, admitting that though it gives him 'honour' so also 'trouble'. Like Doolittle, he recognizes that Baxter does not want to play an authoritarian role in his life, ministry and conscience: 'I will make you if you please a father but not a pope to mee I am resolved to observe a liberty of examining all your resolves yea though you were an angell'. Here there is an echo of Paul's injunction to the Galatians: 'But though we, or an angel from heaven, preach any other gospel unto you than that ye have received, let him be accursed' (1:8). Pinchbecke then asks for the meaning of several phrases of scripture and pleads in 'all your letters be sure to say some thing against idlenesse'. He has heard Baxter is planning to re-issue his 'Aphorismes' and urges that he go ahead, for he had 'gained by them as great as by any booke what ever that I know of'.[54]

In a later letter Pinchbecke thanks Baxter for his 'christian counsell, wherein, by seemeing to passe by idlenesse untoucht you gave it a strong, & sharpe backe blow'. He shares with Baxter some of the problems he faces in his capacity as chaplain to a noble family. It is easy to 'read & pray in private' but to be 'faithfull' in the 'discharge of my duty' when he meets 'with selfe conceited, proud, scornefull, refractorie persons, that are ready to count \mee/ a foole, & a precisian, or a usurper of more authority then is given mee' is far more challenging. Pinchbecke provides a vignette of the way in which Baxter's letter to him enacted a role beyond that of private counsel as he sought to fulfil his more public responsibilities:

> when I received your last letter My Lord, & I were discoursing about chatechizing … upon that occasion I sayd to him My Lord! I have received a letter fro[m] Mr Baxter, wherein he puts mee on very much to duty diligent, & frequent, whereupon he asked me to shew him the letter (that which indeed was my designe) which I presently read to him, so farre as concerned him … \he/ desired mee to write to you, to know your judgement concerning the way of putting that in practise.

Pinchbecke duly refers the matter to Baxter, signing himself off with a variety of alternative role-positions: 'your brother, nursling, or son'.[55]

Baxter's reply has left Pinchbecke with 'no excuse' in implementing the 'unpleasing taske' of catechising the family, which requires him to balance 'plaine instructions or reproofe' with 'language as sweet & my carriage as humble as sincerity will suffer'. The 'humor of our times' has made 'mens eares … very tender'; chaplains thus had to deal tactfully with the landlords who employed them.[56] The position

[53] MS 59.IV.170.

[54] MS 59.VI.155.

[55] MS 59.V.50.

[56] MS 59.VI.146.

could also be rather isolating. Pinchbecke was 'bound to family dutys' and could not go to public meetings abroad for 'my Lord ... knows not well ... what authority they have to keepe them'. The only private meetings are those among Independents 'who will not admitte of mee unlesse I would make my selfe one of theire congregation' which would necessitate 'a breach from others & giveing offence'. As a result he has 'almost noe body' he can speak to or pray with. It is this that makes his correspondence with Baxter so integral to his spiritual health and personal development: 'I do esteeme a cordiall wise faithfull friend that will deale plainely & roundly with mee above all the treasures in the world'. He continues by narrating the struggles of his heart in a manner that is very similar to Doolittle: 'though I pray for a broken heart God seemes not to heare'; he pleads 'oh gaine me if you can an antidote'. Pinchbecke is also anxious, because his parish have asked him to leave and 'My Lord is sicke & as likely to dy as live thus both my crutches breake together'. In a very real sense, he is faced with the loss of employment and financial support; in such a situation a 'cordiall wise faithfull friend' (and potential patron-broker) like Baxter was critical.[57]

In April and May 1654 Pinchbecke sent two letters to Baxter outlining a request he had received from 'the inhabitants of Mashbury in Essex to be theire minister' and asking for his 'wisest counsells how I should carry my selfe so as to be able most effectually to win souls to God & Christ to setle a Church discipline & to joyn in an association or any thing else that may bring glory to God'. Pinchbecke urges his own inexperience and ignorance and the fact that 'God has given' Baxter 'more knowledge & experience' for the good of his Church; how better can this be employed 'than by directing & helping ministers?'[58] He reveals his own physical weakness: 'I am almost nothing but skin & bones & feare that I am in a consumption'. This has prevented him from engaging in 'extraordinarie prayer'. Pinchbecke is also conscious of his failure to do the good he ought to have done in his present position through 'feare & compliance & pride & ambition & sensuality', as he moves to his new employment as a parish minister he desires to be 'more syncere'.[59]

As Neil Keeble has noted, the 1650s raised crucial 'ecclesiological questions' relating to 'the nature, composition, and authority of particular churches ... the significance of baptism, and the criteria for admission to the Lord's Supper ... [These have] an inescapable urgency and particularity to their engagement with fundamental questions of human nature and experience, of human society and the proper extent of its authority over individual consciences' ('Introduction', Keeble and Nuttall 1991, vol. 1, xxvii). The terms in which Pinchbecke asks Baxter for 'counsell' reflects these concerns and uncertainties. The parish of Mashbury consisted of 12 houses; according to the parishioners' own account, it had never had a good minister. Consequently, Pinchbecke writes: 'I know not what to doe I can call none a church

---

[57] MS 59.III.182.

[58] MS 59.V.49.

[59] MS 59.VI.156.

there yet I can \not/ give to any the Lords Supper & for baptisme I can not tell what to thinke well, but I shall incline to the most charitable & mercifull side till I be better satisfyed'.[60] This was a matter closer to Baxter's heart than any other and his response on 5 July 1654 regarding 'the managing of … ministeriall worke' is firmly rooted in his own pastoral experience and practice at Kidderminster.[61] Geoffrey Nuttall has observed that 'certain phrases in it read almost like a rough draft of Baxter's famous work, *The Reformed Pastor* (1656): the style of direct address which marks that book had its origin in personal letters such as this is' (Nuttall 1952, 231–5).

Baxter's letter urges close attention to both public preaching and private pastoral care. God's glory, the salvation of sinners and the unity of the Church are to be Pinchbecke's primary aims. In order to get 'matter & affection' for this work', Baxter outlines what he has found most useful in his own ministry: to study the realities of eternity; to converse privately with each member of his congregation; to study his own heart, which teaches him what others are like; to ponder the weight of the truths he delivers, applying them first to his own heart; to read 'rowsing lively writing' (such as works by Thomas Hooker, William Pinke or William Fenner) and to pray.[62] All Pinchbecke's 'preaching & private dealing' must be done with 'as much love to your hearers as you can … as loveingly as if they were your owne brethren'. In relation to fellow ministers, Baxter suggests that he 'further the work of Unity & association with all your interest & power'. He concludes his detailed outline of pastoral practice with a personal note to Pinchbecke regarding his health: 'Remember also to regard the health of your body; for you will find great use of it for a vigorous serving of God'; this is accompanied by directions for diet and recreation.[63] Similarly, he must take care of his spirits and not 'weaken or damp' them 'by too much sadnes & melancholy'. Perhaps Baxter had picked up the tendency towards depression in Pinchbecke's letters: it is crucial that his congregation are not tempted to think that 'Gods service' will be 'greivious & destructive to their peace & honest mirth but rather let them se in you that godlines is the joyfullest course of life'. This is advice Baxter often urged upon those whose struggle with sin threatened to destroy their joy in God.[64] The closeness of the counsel given here to his own aims at Kidderminster

---

[60] MS 59.VI.156.

[61] MS 59.IV.168.

[62] Francis J. Bremer selects this letter as perhaps 'the most extensive discussion of preparing oneself for the process of researching and writing sermons and other spiritual works' in *Shaping New Englands: Puritan Clergymen in Seventeenth-Century England and New England* (New York: Twayne Publishers, 1995), 23–24.

[63] Baxter gives similar advice in *Gildas Salvianus*: 'Recreation for a student, must be specially for the exercise of his body … they must be as whetting is with the Mower, that is only to be used so far is necessary to his work,' 394. For a thorough exploration of the ways in which medicine impacted on Baxter's life, career and writing style, see Tim Cooper, 'Richard Baxter and his Physicians,' *Social History of Medicine* 20.1 (2007): 1–19.

[64] See for example his counsel to his future wife, Margaret Charlton, soon after her conversion. *Richard Baxter and Margaret Charlton: A Puritan Love Story*, ed. J. T. Wilkinson (London: Allen & Unwin, 1928), 97, 102–3, 128–9.

is seen in the conclusion: 'this is all that in this hast I can give … which little I shall yet find much in the practice myself' (Keeble and Nuttall 1991, vol. 1, 146–8).

For several years after this there is no surviving correspondence. When Pinchbecke next wrote to Baxter on 8 October 1657, he was an assistant pastor to the Presbyterian minister, Thomas Manton, at Covent Garden. It is yet another moment of transition in his ministry and he again turns to Baxter for advice. In this context, the issue is over admission to the Lord's Supper[65]; he also wishes Baxter to provide scriptural proof for his claim in *The Reformed Pastor* that 'a pastor may call his people to an account at any time'.[66] The theological issue is of pressing practical significance, as his Church will soon be holding a meeting about the matter. The following letters deal with these two issues,[67] justifying Keeble's observation that Baxter's correspondence during this period focused on matters of human urgency and the organisation of a society deeply unsettled by the trauma of Civil War and continual experimentation in forms of government. The terms in which Pinchbecke and Baxter discuss the issues also reflect a shift in the configuration of their epistolary relationship very similar to that noted earlier in Baxter's correspondence with Doolittle. On 28 September 1658, after questioning Baxter on the status of the Mosaic law and justification, Pinchbecke identifies himself with Christ's disciples referring to them as an educative model: 'Deare Sir I pray count mee not burthensome but imitate Christ who would solve the doubts of his weake disciples'. Immediately after, however, he urges Baxter to be cautious in his public role: 'that you may never let any errors or unbecomeing expressions fall from your lips or pen for there are many that watch for your falling'. This mutual accountability and Pinchbecke's willingness to invoke it reflects the maturing of his own sense of himself as a pastor and the liberty of his epistolary discourse with Baxter.[68] The necessity for such accountability was something that Baxter himself freely acknowledged: 'I am strongly provoked to blab out any thinge, that I do confidently thinke to be true and weighty…. But as I take your counsaile for caution very thankfully, so I desire you to helpe me by your prayers to obey it'.[69]

4.6 Conclusion

Baxter's correspondence with Lauderdale, Gell, Doolittle and Pinchbecke 'afford[s] telling first-hand evidence of the intellectual and emotional demands made by the aftermath of the Civil War and by the uncertainties of the constitutional and ecclesiastical

[65] The issue of who had the right to receive the sacraments was one of acute controversy at the time, particularly amongst Presbyterians and Congregationalists. See Geoffrey Nuttall, *Visible Saints: The Congregational Way 1640–1660* (Weston Rhyn: Quinta Press, 2001), 135–40.

[66] MS 59.IV.45.

[67] MS 59.IV.45, IV.53, IV.47, IV.54, V.219.

[68] MS 59.IV.49.

[69] MS 59.IV.56.

experiments of the Commonwealth and Protectorate' ('Introduction', Keeble and Nuttall 1991, vol. 1, xxvi–xxvii). Baxter, as both Lamont and Cooper have demonstrated, responded powerfully and prolifically to the violence and disorder of war. The collapse of his health precipitated his writing career in the late 1640s; from this time emerged two books that initiated a spate of correspondence with individuals from all over the British Isles. The Earl of Lauderdale, deeply impressed in his imprisonment by *The Saints' Everlasting Rest,* responded to the deconstruction of his power and privilege by reading, corresponding with, and translating for a clergyman whose authorial identity and wider pastoral authority had been created by the unusual freedom and proclivity of the printing press during the 1650s. The intellectual and emotional exchange that Baxter and Lauderdale shared was maintained entirely by letter; it was tenable only during Lauderdale's enforced leisure and confinement to an enclosed, feminized, spiritual space. Katherine Gell was more willing than Lauderdale to adopt a posture of deference and anxiety construed as stereotypically female. However, she, like Lauderdale, had been prompted to write to Baxter on the basis of his *Saints' Everlasting Rest* convinced of his authority and ability as a pastor and casuist. Unsuccessful though he was ultimately in resolving her concerns, Gell's letters, like Lauderdale's, fostered a network or community of spirituality that helped to establish and enact Baxter's pastoral authority—through the medium of the letter—far beyond the geographical confines of his parish in Kidderminster.

The correspondence with Thomas Doolittle and Abraham Pinchbecke demonstrates even more sharply the tensions that the Interregnum created for young pastors, particularly in defining the nature of the Church and the appropriate administration of the sacraments. It also reveals the remarkable opportunities that this space opened up for experimentation, discussion and reformation. Baxter, in particular, through his active support of young ministerial trainees, his modelling of pastoral care at Kidderminster and his key publication, *The Reformed Pastor*, gained significant authority as both an example and adviser during this period of flux. This pastoral authority and influence was, to a large extent, exercised and maintained through his extensive correspondence network. His exchange with Doolittle helps to close the gap between pastoral presence and the textual traces denoting his epistolary absence. Doolittle's spiritual nurture began as a personal convert under Baxter's ministry at Kidderminster; however, this transitioned to an epistolary relationship when he became a student at Cambridge, supported both financially and emotionally by Baxter's letters. Parochial ministry and personal catechising were not essential to the development of such relationships, though, as Baxter's response to Abraham Pinchbecke's tentative epistolary introduction indicates. In both instances, Baxter's approachability, honest criticism, recommendations for reading and practice, and willingness to debate the theological controversies of the day offered a form of epistolary nurture that enabled the young men to transition from the role of pupils to that of co-workers in the ministry. It was, to a large extent, the unique context of the Interregnum that created Baxter's pastoral authority and enabled him to exercise it in this epistolary form.

References

Black, J.William. 2004. *Reformation pastors: Richard Baxter and the ideal of the reformed pastor*. Milton Keynes: Paternoster.

Cooper, Tim. 2001. *Fear and polemic in seventeenth-century England: Richard Baxter and antinomianism*. Aldershot: Ashgate.

How, James. 2003. *Epistolary spaces: English letter writing from the foundation of the post office to Richardson's Clarissa*. Aldershot: Ashgate.

Keeble, N.H., and Nuttall, G.F. 1991. *Calendar of the correspondence of Richard Baxter*, 2 vols. Oxford: Oxford University Press.

Nuttall, Geoffrey. 1952. Advice to a young minister: By Richard Baxter. *Congregational Quarterly* 30: 231–5.

Chapter 5
Letters and Records of the Dissenting Congregations: David Crosley, Cripplegate and Baptist Church Life

Anne Dunan-Page

5.1 Introduction

On 30th September, 1717, the Church Book of the London Baptist congregation meeting in Curriers' Hall, Cripplegate, recorded the following entry:

> Septem 30 1717: being the publick Church Meeting for decipline affter Sum time was spent in prayer, proseeded to Busness
>
> A Letter being recd from Sister Emerton was presented to the Church & Read, in which she desires her dismission To the Baptised Church at Wantege
>
> Agreed That a Letter of desmission be sent her and Br Skiner is desired to draw it up
>
> A Letter received fom Br Edward Belchamber was Read
>
> A Letter from Sister Katherine Dean was Read
>
> A Letter from the Church of Christ at Exon was presented to the Church and Read, Relateing to Mr Trend, But was not agreable to what we desired it was Resolved that <this> Church will receive Mr Trind into full Communion haveing receid a Satesfactory account of his repentance and humeliation and by virtue of their Letter dated May 26[1]:

This entry reveals that the 'business' of a metropolitan congregation at the beginning of the eighteenth century was transacted to a large extent by letter. Letters

This chapter was written during a period of academic leave granted by the CNRS at the Maison Française d'Oxford in 2010–2011. It could not have been researched without the support of Regent's Park College, Oxford, whose manuscripts I cite with permission. I owe a special debt of gratitude to Emma Walsh and her colleagues, Julian Lock, Emily Burgoyne and Sheila Wood, and for discussion about Crosley, to John Briggs, Stephen Copson and Timothy Whelan.

[1] 'Cripplegate (Curriers Hall) Church Book, 1689–1723', The Angus Library and Archive, Regent's Park College, Oxford, FPC E1, fol. 99v. Thereafter CCB. Interlinear insertions are indicated by < >, editorial interventions by [ ] and deletions by ~~word~~. The original spelling and punctuation have been maintained.

A. Dunan-Page (✉)
Aix Marseille Université, LERMA (EA 853),
29 avenue Robert Schuman, 13621 Aix en Provence, France

IUF
e-mail: anne.page@univ-amu.fr

A. Dunan-Page and C. Prunier (eds.), *Debating the Faith: Religion and Letter Writing in Great Britain, 1550-1800*, International Archives of the History of Ideas 209,
DOI 10.1007/978-94-007-5216-0_5, © Springer Science+Business Media Dordrecht 2013

were received and brought to the attention of the community, letters were 'read', letters were 'draw[n] up', letters were 'sent', letters were 'presented' for the Church to discuss and letters were preserved for future reference. Some were sent from members to the Church, asking to be dismissed. Some were sent to other Churches, either to recommend an applicant for admission or to investigate controversial cases as in this example of a Brother Trend (or Trind) whose repentance for an unnamed offence was deemed unsatisfactory. Once a month, in a meeting dealing with broader issues of discipline, ordained elders and deacons came together with the laity, women with men from the communities of London, Wantage and Exeter. Individual items were carefully examined after a prayer for the harmonious expedition of the day's affairs. Decisions to write letters, and decisions about their contents, were reached collaboratively and letters were signed by (male) members on the authority of the whole Church.

As this entry shows, dissenting letter-writing was not confined to the larger correspondence of a few gifted individuals, writers or ministers. Emerton, Skinner, Belchamber, Dean and Trend are Baptist letter-writers who have left no other trace in history than their mention in their Church records. Yet they all participated in a dissenting epistolary culture that this chapter will attempt to chart through the example of Cripplegate's correspondence with its most controversial pastor, David Crosley.

5.2 Church Records and Epistolarity: The Example of Cripplegate

The letters of 1717 were only summarized, not inserted. The issues raised by Dean or Belchamber were not deemed worthy of inclusion, perhaps because they dealt with a matter not judged sufficiently important or because they were already well-known to the Church. Letters of recommendation were rarely dwelt upon either; they were a staple of congregational discipline, preserving good inter-congregational relations and ensuring that the Church had nothing against an applicant. A conversion narrative was not required of those who had already been in full communion with another Church and a recommendation letter was generally all that was known about a newcomer.

What claims can be made for the value of Church records for the historian of epistolarity if those documents often cross-reference letters without describing or giving the originals, and if letters were perhaps sometimes omitted to avoid transmitting an embarrassing image to posterity? The question raises the broader issue of the records as sources for the history of gathered Churches. Ninety years ago, Wheeler Robinson launched an appeal for denominational history as 'anthropology' and used Baptist records, and the Cripplegate book in particular (whose first folios, or 'book of discipline', he had edited the year before), to show how historians could approach Church life and social history (Wheeler Robinson 1924–1925, 109). Some ninety years after Robinson, historians are especially concerned with the partiality of the Church records and with deceptive patterns in what happens to survive, both of which may limit the value of these documents for the comprehensive project that Robinson envisaged. Baptist records are still buried, for the most part, in archives or lie uncatalogued in private hands; no systematic attempt has been made to identify and study them and they still await their historian such as the colonial and congregationalist materials have found (Cooper 1999; Halcomb 2010). This chapter works towards a study of the surviving early Baptist records, one that pays particular attention not only to their contents but also to their form, and which seeks to explore the ways that letters, among other documents, participated in the writing of a Church's history. Not all letters were as cursorily described as those mentioned above and Cripplegate is a particularly striking example of how they were preserved, understood and used.

When its surviving records open in 1689, this Church had been in existence for over 40 years. It was ministered to by Hanserd Knollys, assisted by Robert Steed, who replaced Knollys when he died in 1691. It changed location many times, especially during the Restoration. In 1689 it was meeting in George Yard, Thames Street, then moved to the Bagnio, Newgate Street, and finally to Curriers' Hall, Cripplegate. It had grown by then into a large congregation, drawing members, according to its first register, from an area beginning at Westminster in the western extreme, then passing along the Strand into Haymarket and 'the Middle of the Citty', reaching beyond the old walls east to Whitechapel, Bishopsgate Street and Wapping, then crossing the river to Southwark.[2] The opening register records the names of 84 women and 41 men, a 2:1 ratio not untypical of other metropolitan Particular Baptist congregations in the eighteenth century (MacDonald 1982, 94–5, 108, 131, 221, 224, 292).

No matter how precise printed Church orders might be concerning the correct procedure for gathering a Church, for selecting and ordaining officers and dealing with cases of discipline, there was no guidance on how Church records should be kept, hence the notable diversity of the materials they contain: registers of members, minutes of meetings, accounts, disciplinary cases, letters, narratives of the gathering of the Churches, controversies and details of ordinations. The terminology, however, was fairly consistent. Church affairs were transcribed in a 'book' often known as 'The Booke of Records'.[3] Congregations paid particular attention to these valuable objects, often taking care to record their purchase and the price paid: 'This Booke was bought by John Lupton at London: and is to keepe in memory such things as are of much Conscernment and of spetiall note hapninge to the baptised

[2] CCB, fol. 1r.

[3] Reading, 1656–1770, photocopy of the original in Berkshire Country Record Office, ref. D/N2 1/1, The Angus Library and Archive, Regent's Park College, Oxford, fol.1.

people and Churchis in Lincolnshire and in speciall for and belonginge to the Congregation at Consby and Tatershall.'[4] In Covent Garden, 'Bror: Price doe buy a booke to enter therein the Act of the Churches and all matters & p[ro]ceedings relating to the said meeting[5]; in Bromsgrove, 'The Church Booke cost 2 sh. 3d day of the 4th moneth 1670'[6]; and in Slapton, on 13th June 1690, the deacon, paid one shilling 'for this Booke'.[7]

The contents of each book were left to the judgement of the pastor. One of the minsters most forthcoming on this subject is Isaac Gardner, the pastor of Hamsterley and Cold Knowley:

> N.B. This Book Came to my hand soon after Bror [William] Carrs Death & I find their has been Omitted writeing any remarkable passages in it, since ye year. 1731./2. It belongs to ye Church of Christ, Baptised on ye Profession of Faith, meeting at Hamsterly & Coldrowly in the County of Durham: & is Desighned for ye Churches Book, wherein ye Minister or Ruleing Elder or Pastor is to set downe some of ye most remarkable Occurrances & Transactions of ye Church, not only for ye more Orderly Observeing of ye things Concluded on for ye time present, but also for ye Direction and Comfort of ye Generations to Come, into whose hands, thro' ye providence of God, it may fall. Isaac Garner.[8]

The book belonged to the Church as a collective body. According to Gardner, it was meant to preserve the memory and history of the Church, a conception shared, for instance, by the Reading community whose book is called 'A Book of Remembrance'.[9] The congregational Church at Axminster used exactly the same terms:

> Shall kingdoms and commonwealths have their chronicles, civil courts their rolls and records? Shall tradesmen keep their books of accounts, lawyers their books of precedents, physicians their collections of experiments, and travellers their journals? And shall not the churches of Christ have their registers and books of remembrance wherein they may record their church transactions and the various dealings of God with them?[10]

If a Church book recorded 'remarkable Occurrances and Transactions', most congregations used it as a minute book, with abridged transactions of meetings, evidence of the movement of members and, occasionally, financial accounts.

---

[4] Coningsby and Tattershall, 1654–1728, The Angus Library and Archive, Regent's Park College, Oxford, fol. 3.

[5] 'The Church Book att Covent Garden Anno: 1691 to Anno: 1699', The Angus Library and Archive, Regent's Park College, Oxford, fol. 4.

[6] Bromsgrove, volume 1, 1670–1715, The Angus Library and Archive, Regent's Park College, Oxford, fol. 3.

[7] Slapton, The Angus Library and Archive, Regent's Park College, Oxford, fol. 7.

[8] '1651–1832. Records and Letters Relative to the Baptist Church at Hexham, from Oct 1651 to July 1680. Followed by Records of the Church at Hamsterly in the County of Durham', The Angus Library and Archive, Regent's Park College, Oxford, fol. 17.

[9] Reading, fol. 1.

[10] *The Axminster Ecclesiastica, 1660–1698* (1874), ed. K. W. H. Howard (Ossett: Gospel Tidings Publications, 1976), 2.

Any uncertainty as to whether Church books should record the extraordinary, the mundane, or both were solved in the largest and most prosperous congregations of London by keeping several books, depending on the nature of the material. In February 1693, Luke Leader of Maze Pond was instructed to buy a separate book for the accounts and poor-money, as he was keeping, in parallel, 'the Church Book to wright therein the Churches proceedings'.[11] Devonshire Square had separate registers of members and account books.[12] Since no financial accounts appear in the Cripplegate records, that community may well have adopted the same solution.

For the period 1689–1723, Cripplegate did not only have a Church book, and, perhaps, a separate account book; it also had a 'minute' book which has not survived but is mentioned several times in the Church book. In January, 1703, it was decided that James Newton would be paid thirty shilling every three months for visiting the members who had failed to contribute to the maintenance of the minister and for 'his writing ye Minutes, and other concerns of ye Church'.[13] Entrusting the minutes to a paid lay member, while the pastor kept the Church book, may suggest that the latter document had precedence over the former and was a choice place for 'remarkable Occurrences' only. And yet, Cripplegate never clarified the relationship between its two volumes.

The Cripplegate Church book begins with a calendar of its first folios, compiled in 1707, followed by an undated register of members. It continues for 11 years as what Wheeler Robinson called a 'discipline' book (Wheeler Robinson 1922–1923). There are entries concerning admonitions and excommunications, consistently written in Robert Steed's small and neat hand, from 26 February 1689 to 14 February 1700, with insertions of later material since Steed never felt the need to fill in the whole folio. It is not entirely accurate, however, to describe that first decade of records as entirely 'disciplinary' since it also contains Steed's version of the departure of some members to found a new Church. These, including Luke Leader and Edward Sandford, had temporarily joined Cripplegate after they had left Benjamin Keach's congregation in Horseleydown over the question of hymn singing. When the time came to embody their own Church, a controversy erupted in Cripplegate, Steed deeming them not mature enough to 'sett downe by themselves as à [sic] distinct church'.[14] This prompted the departure of Steed's assistant, Richard Claridge, who had sided with Leader and Sandford.

From Steed's death in 1700, the Church began recording what Isaac Gardner would indeed have called 'remarkable' events. The schism of Richard Paine, who took exception to the Church's offer of the pastorate to David Crosley, was followed by the ordination of Crosley on 12 February 1703 in the presence of the London elders Joseph Stennett, Richard Adams and John Pigott, and of some deacons. Shortly after followed amended 'Articles of the Church' in 1705, followed by the

---

[11] 'Maze Pond Church Book, Folio (1691–1708)', The Angus Library and Archive, Regent's Park College, Oxford, fol. 93.

[12] Guildhall Library MS 20228/1A and MS 20230.

[13] CCB, fol. 23v.

[14] CCB, fol. 5r.

case of a converted Quaker couple who sought admission while professing 'to keep ye 7th day'. Finally came the renewal of the covenant, and the decision, conducted in the form of letters, not to attend the 1706 meeting of the London Association on the grounds that it admitted Seventh-Day and Arminian congregations. Beside these exceptional events, however, the book continues to record routine matters, such as the admission of new members, disciplinary measures, and decisions taken at Church meetings. The problem of an overlap between Church book and minute book, in other words, was never successfully resolved, as was perhaps not seen as a problem at all. From 1711 onward, the distinction became even more tenuous and the Church book seems to fuse with the minutes as most entries record the decisions taken at the monthly meetings.

Cripplegate, therefore, differs from other Churches whose records have survived in the way it experimented with the contents of its book and minutes. The existence of the minute book, I would suggest, is partly responsible from the transformation of the Church book, from May 1707 to May 1711, into a lengthy account of Cripplegate's proceedings against David Crosley, in which we find the correspondence between Cripplegate and its fallen pastor, and between Cripplegate and provincial Churches. The inclusion of epistles is certainly not unique in Church records, as we have seen in the case of recommendation letters; Michael Davies has recently drawn attention, in a pioneering essay, to the letters of the Bedford congregation (Davies 2009); Cripplegate, however, is unique in the way it preserved controversial letters for the best part of 4 years, transcribing them, amending them, and using them as legal documents in one of the biggest scandals to afflict the early eighteenth-century dissenting community.

5.3 David Crosley

Weakened by the separation of the former Horseleydown members, by the departure of Paine and its supporters together with Richard Claridge, Cripplegate was without a pastor between 1700 and 1703 when David Crosley was formally ordained, having been dismissed from Tottlebank, the congregation he ministered to in Yorkshire.

Crosley (1669–1744) was not an unwise choice for a large and ancient London congregation. Born in Heptonstall, near Todmorden, on the border of Yorkshire and Lancashire, he became a Baptist in 1692 and was baptized in the Worcestershire congregation of Bromsgrove. He then briefly settled in Barnoldswick before accepting the Tottlebank pastorate in February 1696, against the wishes of an association of local Churches held in Barnoldswick (Whitley 1913, 77–80).[15] 'A name rich with

[15] For Crosley's adoption of Baptist views through the agency of 'one woman of any accompt', see 'The Copies of Some Christian Letters', Greater Manchester County Record Office, William Farrer Collection L1/43, fols [173–8]. Thereafter Notebook. The foliation is inferred. I am quoting from the manuscript rather than Frederick Overend's at times inaccurate transcription of Crosley's letters.

blessed memories' (Lewis 1893, 4), Crosley is revered in Baptist historiography for his missionary travels throughout Northern England. From a base in the forest of Rossendale, he and his elder cousin William Mitchell established 'a loosely-organized circuit' of preaching stations which sometimes matured into gathered Baptist Churches (Blomfield 1912, 73). Whereas Mitchell was apparently content to operate in the North, and was indeed Tottlebank's first choice for a minister, Crosley's nature predisposed him to wanderings and he had already embarked on several preaching tours in the Midlands and the South of England before arriving in Cripplegate.[16]

There is no dearth of information about David Crosley, beginning with James Hargreaves who dedicated a long appendix to Crosley's works (Hargreaves 1816). Most of what we know about him, however, has not been revised since the early twentieth century, and either concentrates on the way the Northern Churches operated or is mainly biographical.[17] The story of David Crosley's most difficult moments, between 1708 and 1711, needs to be retold through the Cripplegate correspondence, not only for what it reveals about the man himself but also about the relationship between provincial and metropolitan congregations and the nature of the Baptist ministry.

I am here chiefly concerned with the letters to and from Crosley and provincial Churches preserved in Cripplegate's Church book. But David Crosley was a productive letter writer. None of his holograph letters seem to have survived although a representative sample were transcribed in a 1692 notebook together with those of his Rossendale friend John Moore, later minister at Northampton. Frederick Overend published an account of the Rossendale Churches based on 17 letters taken from Moore's manuscript (Overend 1912). This belonged to the Lancashire antiquarian William Farrer and was purchased from his widow by the Greater Manchester County Record Office, where it remains today. Item L1/43 in the ocean of Farrer's notes and papers, has not been examined since Overend. It comprises some 500 folios with unregistered letters to and from Mitchell and Crosley. When and where necessary to illuminate the Cripplegate events, I will therefore draw from the letters of the Moore manuscript while leaving a fuller study to appear elsewhere.

[16] See for instance, Notebook, fols [173–8, 181–2, 191–5].

[17] The main contemporary sources of information on Crosley's life are, James S. Hardman, 'David Crosley, the Pioneer Rossendale Baptist', in *Four Articles by James S. Hardman*, published for the Sion Baptist Church (Cloughfold, 1947), 3–13; Ian Sellers (ed.), *Our Heritage: The Baptists of Yorkshire, Lancashire and Cheshire, 1647–1987* (Leeds: The Yorkshire Baptist Association and the Lancashire and Cheschire Baptist Association, 1987); B.A. Ramsbottom, *The Puritan Samson: The Life of David Crosley, 1669–1744* (Harpenden: Gospel Standard Trust, 1991); S.L. Copson, *Association Life of the Particular Baptists of Northern England, 1699–1732*, English Baptist Records 3 (Baptist Historical Society, 1991); M.F. Thomas, *Tottlebank Baptist Church, 1669–1699* (1999, privately printed); J.H.Y. Briggs, 'Crosley, David (1669/70–1744)', *Oxford Dictionary of National Biography*, Oxford University Press, 2004 [http://www.oxforddnb.com/view/article/6790, accessed 28 April 2011].

5.4 The Seventh Commandment

The honeymoon between Cripplegate and its provincial minister was short-lived. On 7 May 1708, at a Church meeting, four brothers were dispatched to 'search out ye Truth of ye scandalous reports on our Elder'.[18] A few months later, the Church opened proceedings against Crosley and transcribed them in the Church book under the title, 'A faithfull Narrative of the proceedings of severall Brethren, and of this Church of Jesus Christ—against Mr David Crosley, their late pastor, from the beginning of decemb: 1707 to the 14 of aug 1709'. It is here impossible to rehearse all the details of the proceedings, but I will concentrate on one especially scandalous charge against Crosley: his behaviour towards women.

Crosley was a heavy drinker, as the Church well knew, but when his excesses began to attract the attention of 'the Elders of the Baptized Churches with whom he had communion', some Cripplegate members thought it was time to take action.[19] Crosley was first heard and admonished privately by some 'Christian Brethren', before being brought before the whole Church. He repented, promised to amend and matters were dropped for a time, but he proved unable to remain sober. Excessive drinking was not, however, Crosley's only moral failure: when in his cups, he had apparently behaved 'immodestly' towards at least three women: the niece of one of the Church members named Hester Hannis, Susan Emerton (who sought a dismission in 1717) and an unidentified woman at The Three Daggers tavern. When Crosley was finally spotted in the company of a reputed prostitute, he invented a story to justify his dealings with the woman and, when exposed, had no choice but to confess his lies. Nonetheless, he never acknowledged that he had ever acted 'immodestly'. Crosley promptly returned to Tottlebank, via Colne in Lancashire, leaving the Cripplegate elders to manage the final stages of the procedure *in absentia*. He was finally excommunicated for 'drunckeness', 'immodest behavior towards women, bordering on the breach of the 7th Commandment' and 'Lying'.[20] For the following 2 years, Crosley corresponded with Cripplegate to convince them to restore him and then dismiss him properly so that he could served in Tottlebank again.

No Baptist ignored what had to be done in cases of offences, for discipline was duly rehearsed in Church orders and practiced in monthly meetings. If the nature of the offence was private, the offender would be admonished privately, asked to answer the charges brought by at least two witnesses and matters would be dropped if repentance were deemed to be sincere (Renihan 2008, 56–57). In case of 'obstinacy', the case would be brought before the whole Church. The unrepentant offender could be suspended from communion in the first instance and failure to amend would result into excommunication. He or she could be 'restored' if their repentance was convincing.

---

[18] CCB, fol. 30v.

[19] CCB, fol. 32r.

[20] CCB, fol. 36r.

Cripplegate was almost obsessively careful in respecting this procedure, believing it could 'affect all the Baptised Christians in citty and countrey' if it were seen either to act precipitately or to delay matters unduly.[21] Although Crosley was dealt with in the same way as any other offending brother or sister, his ministerial position rendered matters open to public scrutiny. As a result, every precaution was taken to gather evidence. For instance, in order to prove that Crosley had never visited a horse dealer on Tower Hill but was there solely on purpose of seeing a prostitute, Church members were despatched to the area with Crosley's letter in hand and 'went from dorr to dore & inquired of ye houskeepers of all ye houses in all ye passages on both sides of ye way'.[22] Witnesses were heard, carefully examined, and asked to depose verbally or in writing, under oath. One of them was the young Hester Hannis, the niece of Matthew Tindall. Hester's charges against Crosley amounted to two: first, 'kissing her immodestly several times', to which he answered that he 'might', when drunk, have kissed her 'longer, or harder then was vseuall'; second (which he entirely denied), 'unchast behavior'.[23] In a private meeting Hester

> solemnly declare[d] that Br. Crosley did severall times behave himself very immodestly to her, in kissing her severall times, and taking her on his knee, and blowing out the Candle, with another wanton, Rude, behaviour, that her Moddesty would not suffer her to declare, to the said Brethren, nor otherways but when her vnckle asked her the question, she answered in the affermitive, and being asked by Br Adams, if she could take an oath, if called to it to what she had now affermed, her answer was she could doe it with a very safe Conscience.[24]

This has to be one of the most telling silences in Baptist records. We are left to imagine what *the* question was, a question asked by a close relative, and meant to elicit an answer without forcing Hester to name the facts.

And still, her testimony was judged insufficient in the absence of a second witness, and when confronted with her, Crosley confessed nothing. Two of Crosley's female supporters, Sister Hannah Wall and Sister Hurst, then cast aspersions upon Hannis's character:

> for thay accused [Hester Hannis], of being a drunckard, lyer, and theif, soe that it caused a great Clamour, and disorder, in the Church, and great Reflections one upon another, because some adheared to Hester Hannises accussaction, Rather then to Br Crosleys defence.[25]

At this stage, Cripplegate was clearly on the verge of splitting over Crosley. The Church would not reach a unanimous decision until a few weeks before his excommunication, when it could be proved beyond any doubt that he had lied to them.

Having failed to produce two witnesses or to secure a confession, Crosley's accusers were left to alter slightly the terms of the law and bring a series of women to corroborate Hester's sayings, which would amount to more than one witness, for

[21] CCB, fol. 34r.

[22] CCB, fol. 51r.

[23] CCB, fol. 33r.

[24] CCB, fol. 33r.

[25] CCB, fol. 33v.

'Respecting Hester Hannis and considering the nature of the crime … it was very improbable, if not impossible, such an offence should have two wittnesses'.[26]

Both adultery and the absence of consent from Crosley's alleged victims were at issue here. All the other deponents whose narratives and letters were included in the Church book stressed (or were made to stress) the same point. Susan Emerton, for instance, testified that:

> when she was walking Mr Crosley came & had a little discours with her yt if she should be out of place att any time she should be welcome to his hous & kist her putting his tongue into her mouth wch put her much out of Countenance when ever she saw him.[27]

One Howell Jones was brought in to narrate what he had seen at The Three Daggers and again the question of consent was raised:

> that he being drinking at the 3 daggers was bid by the maid to observe a passage, soe he looked through a Chink, and saw Br Crosleys face, as close to a womans, as ever his was to his wife, with his armes about her waist, being both standing, afterwards, called for more drink and gave her, and put his hand about her neck, and the other as low as her waist, but saw no further act, only, the woman seemed to withdraw her self from him as being uneasey.[28]

A later account has a slightly different—and decisive—version of the last phrase: 'but I saw no immodesty acted'.[29] When Crosley was then observed on Tower Hill with a prostitute, he escaped north in the spring of 1709.

5.5 Letters and the Law

Before examining how Crosley and Cripplegate conducted this second phase of their relationship, it is necessary to rehearse the status of the letters in the Church records, which points in opposite directions: a particular attention to the legal value of the written documents, combined with a lack of care in their transcription.

Letters in Church records are transcripts of originals and Cripplegate's correspondence is no exception. As we have seen, some of the letters the Church received were judged sufficiently important to be copied into the Church book or minute book, while others were not. It is not clear whether the originals were kept or destroyed, or who had the authority to decide. In the case of Cripplegate, given the extraordinary nature of the events, we have more information than usual on the fate of the letters: both Crosley and the Church had apparently agreed that the originals should be deposited with the deacons, as well as other written transactions, regarless of whether they were included in the Church book.[30] Letter-keeping was not a

[26] CCB, fol. 33r.

[27] CCB, fols 49r, 48v.

[28] CCB, fol. 35r.

[29] CCB, fol. 48v.

[30] CCB, fol. 35v.

codified role for deacons and it is therefore difficult to ascertain whether Cripplegate was simply improvising under pressure or whether deacons were always implicitly trusted as safe guardians of original documents, as they were trusted as safe guardians of the Church's money.

Cripplegate was paying attention to its letters. Originals were preserved, copies of important items were made, catalogues were compiled to make the collection searcheable as, for instance, a complete 'Index to the Following Transactions and Letters' compiled after 1711. It records the folios of 58 listed items, the dates of the letters, the identity of their senders and sometimes the place from which they were sent.[31] Despite this seeming care, the reliability of the Cripplegate correspondence is not to be taken for granted, given the nature of the records. First of all, for the whole duration of the controversy with Crosley, the Church book was abandoned, the congregation probably relying on the minute book only, in the absence of a pastor. From 1708 to 1711, therefore, the entries are not contemporary with the stories they tell. They were copied *a posteriori*, either from memory, from the originals in the deacons' possession, or were perhaps transcripts of transcripts if an item was copied first into the minute book.

Most letters are dated but it is impossible to determine when they were copied into the Church book. After the 'narrative' of the proceedings (necessarily compiled after August 1709), we find material in another hand relating to the year 1702, a second account of Crosley's ordination, several recommendation letters dated 1690, 1705 and 1708. The compilers (there are at least three different hands) first used the recto of the folios only, but added material onto the rectos, making it necessary to insert indications at the bottom of a page such as 'red now in page 47' or 'look now to page 49' not to interrupt one particular item.[32] Letters to and from Crosley and Tottlebank are intertwined with depositions that are mentioned, but not transcribed in full, in the narrative of the proceedings (Hester Hannis, Susan Emerton, Howell Jones…), resulting in different versions of the same testimonies.

Second, there is no evidence that the letters were recognized as having a particular generic identity. When dealing with letters in Church records, it is vital to keep in mind that they were documents embedded in the legal proceedings of a congregation, and valued as legal evidence. Crosley's letters were not judged to be different in nature from recommendation letters or spiritual letters but all functioned as 'transactions' of the Church to maintain order and displine. All had the same legal status, all were parts of the 'acts' of the Church, whether they recommended a brother, admonished another, supplemented an oral testimony or vindicated some charge. There is no indication, as the miscellaneous index shows, that they were given separate status among 'recommendations', 'accounts', 'narratives', 'depositions', 'answers', 'notes', even 'treatises'.[33]

---

[31] CCB, fols 37r–37v.

[32] CCB, fols 46r, 48r.

[33] CCB, fols 37r–37v.

Third, letters did not constitute a satisfactory mode of communication. Whether an applicant was giving a conversion narrative to be admitted into the Church, whether a member was admonished, answered charges or repented for an offence, the Church systematically gave primacy to the spoken word. Letters were seen, at best, as supplementary material, at worst as a testimony of the person's guilt, when his or her shame or stubborness prevented physical encounters. Crosley was well aware of this and he implicitely accused Cripplegate of forcing him to adopt an unsatisfactory mode of communication:

> With a very deep & penitential concern I write & desire these lines may be read & sedately weighed among you & I chuse this method (this opportunity) rather yn to be personally<present>because ye discomposure & sorrow wch now drink up my spirrit under ye consideration of ye manifold rebukes wch on a sudden like an armed man Come upon me…[34]

It is only out of 'duty' and 'necessity', when forced to seek a restoration from Cripplegate, that he would overcome his reluctance to write, '<a>sence of either of wronging my self or offending of you still keps my pen however now at last a sence of duty Edged with necessitie has made me resolve to lay my Case and heart open before you'.[35]

Finally, the letters transcribed in the Church book bear the trace of the haste and carelessness of the transcribers. In the most severe cases, the records candidly acknowledge that they are but 'an Imperfect copy' and that some transcriptions are redundant: 'This letter is now by mistake twice Enterd'.[36] Whereas the 'narrative' had been written with the utmost care, some sentences in many letters do not make any sense at all. The folios were therefore corrected by another hand, trying to reddress the most basic mistakes (but not necessarily helping the modern reader): there are interlinear insertions when one or more words are missing, or a whole sentence needs to be rectified; deletions, corrections of the spelling and scribbled strips of paper secured by pins were even added in the middle of a folio when space was lacking for amendments. Somebody evidently thought that the Cripplegate letters were almost useless in their original state, if not plainly wrong, as a guide to the events and that the original scribes had failed in their duty of preservation.

Letters and Church records stood therefore in an uneasy relationship to one another, no doubt because of the multiplicity of versions: the original, the transcription in the minute book and the transcription in the Church book. Letters served as material proofs when a controversy arose but there is no evidence that their nature was recognized as different from that of other written transactions. Despite the seeming care taken to preserve the originals and transcribe their contents, and despite later efforts to amend and catalogued letters, what is recorded in the Church book was sometimes selected, sometimes transcribed in full, sometimes integrated

[34] CCB, fol. 46r.

[35] CCB, fol. 64r.

[36] CCB, fols 50r, fol. 51v.

in partial narratives, sometimes left with no commentary, testifying to an uncertainty over the precise status of the documents and a refusal to lessen the preeminence of the spoken word.

5.6 A Wounded Spirit?

David Crosley acknowledged his drinking and lying but not adultery, 'for there is nothing I more abominate, and abhor'.[37] Whether or not he was guilty is a moot point. The testimonies of the three London women, further reports of 'scandalous' behaviour in the North in 1719 and 1736 (Blomfield 1912, 87–86), and a letter mentioning how his engagement with a widow had been broken off because he was accused of taking advantage of his charismatic preaching, point in the first direction.[38] On the other hand, the formulation of Cripplegate is cautious, '*bordering on the breach* of the seventh commandment' (my emphasis) and no consensus was ever reached as to the precise nature of Crosley's sexual offences. One of his chief accusers, the uncle of Hester Hannis, would later seek a dismission from Cripplegate, softened his accusations and acknowledged he had been 'taken with [Crosley]'s spirit as to ye Doctrines of ye gospell & wth his person as to ye loving free & pleasantness of his Conversation', although he remained convinced of his guilt.[39]

The historian's sole concerns are with the spiritual and practical arguments of the correspondence. Crosley's confession came with accusations of Cripplegate's 'ill nature'. They had been 'wrong in ye Management & Execution of things' and 'groundless Iealosies & Insinuations' had forced him to lie.[40] Such a course of action was incautious but Crosley curiously never altered it, constantly attacking the way his (public) excommunication had been mismanaged through the unnamed (personal) 'Iealosies' of his accusers. In the end, Crosley would recognize his excommunication as a proper act of the Church yet without ceasing to berate Cripplegate for its lack of charity:

> Tis true your proceedings wth me has very sencibly affected me and proved a burden very heavy to be born but since tis ye Act of a Church & of such a Church I have owned the Authority of Christ with you and not disputed as in many things I Iustly might but submitted as a poor sinner aught But now how is it that I have not so much as one line from you to Exhort comfort or support me or in ye least to pave the way for my restoration.[41]

Crosley adopted several strategies. He first mounted emotional epistolary defences, displaying the self-lacerating rhetoric of the penitent sinner which he knew might soften Cripplegate and encourage them to reverse their sentence.

---

[37] CCB, fol. 35v.

[38] Notebook, fols [312–16].

[39] CCB, fol. 61v.

[40] CCB, fol. 63r.

[41] CCB, fol. 63v.

He compared himself to Job, Jonah and Jeremiah, using variations on Proverbs 18:14: ‘yt I may not altogether sink in my spirit whom am your wounded distressed & Sorrowful Brother David Crosley’, ‘A wounded spirit who can bear DC’.[42] A broken man assailed by ‘mallancholly’ on the brink of ‘Desparation’—but still a child of God—he asked for prayers and pity instead of harsh rebukes.[43] Whether we choose to believe him or not, there is no doubt that Crosley had been afflicted early on with strong temptations, as testified in a letter he wrote, aged 18, to his cousin Mitchell, ‘I left desolate & cannot deliver myself; & so fiercely doth my Souls adversaries many times assault me, that I am even overwhelmed thereby’[44] or, a few years later, to his friend John Moore, ‘I rest a poor worm in my selfe, & sore distressed, troubled & afflicted in ye flesh’.[45]

Crosley was quick to suggest that the Tempter had taken advantage of his ‘naturall disposition’: ‘o yea Enimies of souls & of my soul I am sure now he hath over come me’.[46] However, he was no less reluctant to lessen his own responsibility invoking, for instance, his ‘laxness as to close attendance on study & private duty’[47] and the way he had underestimated the ‘vast difference betwixt city and Countrye’.[48]

Guilty and yet a victim of men’s prejudice, of Satan’s snares, and of London’s perverting influence, the pastor then turned away from spiritual suffering to practical compromises, still balancing the minutiae of Church government with impassioned rhetoric. Crosley believed it was his duty to give lessons in Apostolic government to his former Church: ‘Under the severest discipline of ye primitives times none were kept yt I can remember in ye place of Pennitents under Censure above 3 years’.[49] He first suggested he could resign and ‘retire among [his] old acquaintance in ye north whereby ye desire as well as want of ye people is great.’ When that failed, he thought to obtain ‘a discharge tho without recommendations’.[50]

Unable to convince Cripplegate that God had given him repentance or that he should be allowed to resign, Crosley argued that ‘hard usage’, excommunication and a refusal to restore him would prevent him from being ‘serviceable’ and ‘usefull’ elsewhere. By adopting an inflexible attitude, Cripplegate was not simply dealing unjustly with him: it endangered the spread of evangelisation of the Northern counties,

[42] CCB, fol. 46v.
[43] CCB, fol. 53r.
[44] Notebook, fol. [97].
[45] Notebook, fol. [248].
[46] CCB, fol. 52v.
[47] CCB, fol. 56r.
[48] CCB, fol. 56r.
[49] CCB, fol. 64r.
[50] CCB, fols 46v, 48v.

> Butt I should yett have sat still in silence and have wayted for your own Bowells to have moved the waters but that ye crys of ye people of God with & about me here for & after ye Ordinances of Christ forces me I have Urged ye Circumstances I am in & begd them to cast their ey upon some other but they whollie decline it I have told them I greatly Hesitated whither ever I should engage in those sacred services any more but they will not Endure ye hearing of it. They are still at me to know what they can do to facilitate my regular dismission without which I am not willing to act I tell them I hope a little while will sattisfie them. they add a many hand at ye Churches door desirous of admision but are not willing to proceed till they se me first fixt.[51]

For all its subtle balance of technicalities ('facilitate a regular dismission', 'proceed', 'fixt') and feelings ('Bowels', 'ye Crys of ye people of God', 'Endure'), Crosley's prose again failed to move Cripplegate. In 1736, among a series of 'Christian councels', Crosley warned his readers against the perils of keeping company with swearers, unjust dealers and drunkards, but also 'censorious' and 'uncharitable' people, reminiscent of his Cripplegate accusers: 'no Relation or Tye so sacred, but they will violate it; no Person or Office so publick or useful but they will readily expose both to gratify their own pevish and censorious Humour; as if their own Reputation and Interest never rose so high or stood so firm, as when founded on the Ruins of another'.[52] This is how David Crosley still felt some 30 years after leaving London.

Cripplegate had not been entirely insensitive to Crosley's arguments. Their first letter to Tottlebank was meant to inform them that Crosley had been cast out, 'till God give him repentance', but without mentioning the specific charges, using instead what they called 'Generall terms'.[53] Unconvinced by Crosley's defence, they could not endanger London for Yorkshire, 'we dare not revert ye Order of ye Gospell here for the sake of ye Gospell in your parts'.[54] Tottlebank was less than satisfied and the Church embarked on a series of vitricolic epistles, considered 'plain abuse' by Cripplegate, to press the latter to dismiss their pastor.

The exchanges between Crosley and Cripplegate and between Cripplegate and Tottlebank reveal the special relationships between a pastor and his congregations and, beyond, different conceptions of the nature of the ministry. After Crosley's departure to London, Tottlebank had continued to feel, wrongly according to Cripplegate, that they 'caried a supposed prior right and Interest'[55]:

> His work should have <inspired you with> more Candor & tenderness to <ward> him in ye day of his rebuke suffer us to ~~tell~~ represent <to> you ye barbarity of yt part of yor conduct even as <it were> tearing our pastor as we esteemed him out of our bosom & slaying him you know wt sence before our eyes & while you ~~would~~ could not hinder Iust providence from making restitution in returning him to us again yet wod continue ye proofs of yor ill nature in making him as uncapeble as possibly you could be being <either> Comfortable or usefull among us.[56]

---

[51] CCB, fol. 64r.

[52] 'Plain and Honest Directions, and Christian Counsels', in N.T., *The Old Man's Legacy to his Daughters* (1736), 91.

[53] CCB, fol. 36v.

[54] CCB, fol. 66v.

[55] CCB, fol. 57r.

[56] CCB, fol. 54r.

For Cripplegate, such accusation of 'barbarity' was no mere rhetorical hyperbole. Tottlebank implied that they had called David Crosley without respecting the proper procedure. And again, letters would serve as evidence of the legality of Cripplegate's conduct:

> if our Conduct in our first calling him was not right it is chargable on him and not on us for our first request was to know his state and brother crosely positively assarted his free state from those ~~of~~ gospel bonds or tys that might hinder his remove ye letters we have carefully preserved and afterwards he declared ye same before many elders and eminant Christian that left no rome for us to suspect ye truth thereof.[57]

For Crosley and his friends in Tottlebank, human weakness was a spiritual advantage, Crosley being more able to cure wounded souls, having himself experienced strong temptations. Pastoral letters to various correspondents in the Moore manuscript confirm this point. Tottlebank did not entirely disculpate Crosley but the Church was convinced that his letters, as well his sermons, displayed the proper repentance of a broken sinner:

> for our part we went only to represent wt we have gathered ~~to~~ <from> his ~~sermonces~~ <sermons & particularly> from his private discourses Littrs & espeacially to those whom he was more intimate with <wherein> he as told us ~~the means~~ of his nature wt strugles he had with ye body of sin how hard set he was to bear up agst ye stream of corruption & temptatiom & wt fear he had least ye enimies <should> prevaile this & much more we met with from him yet for all this we are not ashmd to say we Loved him not ye worse <for it> becawse he better knew <himselfe &> how to humble himselfe & how to spake to others so as to be ye more usefull.[58]

Telling of his nature, his struggles, his fears, his stream of temptations was precisely what Crosley had done, to no avail, in his letters to Cripplegate but the smaller, Northern, community spoke of 'love' instead of procedures, of intimate exchanges with friends, instead of public epistolary vindications.

Finally, the success of a minister was measured less according to his moral conduct than according to his usufulness in making converts in the 'interest of religion'.[59] As a consequence, the ability of a congregation to maintain his minister, which depended on the size and wealth of its membership, was judged to be of God's ordering. When, in 1710, Cripplegate poached Joseph Matthew from Grittleton (although he was not ordained after 3 years of trial), the small congregation consented to let him go on the grounds that they were too few and too mean to retain him, thereby signing a warrant for their own disappearance. Keeping him would have been a 'dishonour to god & a blot to o[u]r holy profession', they wrote to Cripplegate.[60] The economic argument was subservient to the godly design that directed a minister where his efforts would be rewarded by an increase in the community. Crosley himself repeatedly put this into practice: in 1695, he accepted the

---

57 CCB, fols 56v–57r.

58 CCB, fol. 59r.

59 CCB, fol. 53v.

60 CCB, fol. 60v.

pastorate at Tottlebank against the advice of an association of Churches; in 1736 he was accused of maintaining an 'irregular way to increase his members ... by all the artifices he can use' (letter of John Marshall quoted in Whitley 1913, 95). Where conversion was at stake, Crosley never hesitated to put the spiritual interest of sinners first.

According to this numerical logic, Crosley had many an advantage. Even Cripplegate recognized that the congregation had flourished under his pastorate. He was not, however, simply a gifted evangelizer, but a minister deeply concerned with proper Church government. Long before he settled in London, his cousin Mitchell had asked him to pronounce over issues of discipline.[61] In 'Christian Exhortation to Church-Fellowship', transcribed in the Moore manuscript, he defines his ministerial mission as feeding the neighbouring people 'sound Doctrine' and 'endeavour[ing] to promote amongst you such a Gospel-Order & Christian Fellowship as may most tend to your furtherance in ye Gospel'.[62] Even the most incisive commentators, such as Murdina MacDonald, have failed to realize that the ordination by the laying-on of hands of neighbouring London ministers was proposed to Cripplegate by Crosley himself, a 'method' of ordination he was keen to promote: 'the method proposed by Br Crosly and agreed to by the Elders (viz) Br Adams Br Pigot & Br Stennet in ye ordination of Br Crosley feb 12 702'.[63] The picture of William Mitchell as the steady builder of Churches and of David Crosley as the 'rolling stone' or 'erratic individual' of their evangelical partnership (Hayden 2005, 102; Blomfield 1912, 78) should be nuanced in the light of Crosley's never failing interest in Church government.

For Tottlebank, condemning Crosley to forsake his ministry ran counter to the will of God, because of his pastoral success and the necessity to evangelize the North, many hands beeing at 'ye Churches door'. For Cripplegate, Crosley's lack of repentance was unacceptable and prevented his restoration and proper dismission. The nature of his offence, as often in cases of excommunications, mattered less than his obstinacy. None of the parties would ever reconcile.

5.7 Conclusion

The Cripplegate correspondence contains one of the best-documented scandals among early eighteenth-century gathered Churches and yet the formal relationship between letters and Church records had never been examined. Replacing the dissenting letters among their legal context alert us to the fact that epistolary exchanges could as readily sever the network of Churches as reinforced them. A vital element in the 'organizational response' of nonconformity to both persecution and toleration, as Richard Greaves termed it several decade ago (Greaves 1975), letters could

[61] Notebook, fols [437–40].

[62] Notebook, fols [238–44].

[63] CCB, fol. 38r.

also initiate and fuel controversies over Church government never to be healed. I have barely scratched the surface of David Crosley's extraordinary career: the way his printed works interacted with his correspondence or the richness of the Moore manuscript. Yet the exchanges preserved at Cripplegate illustrate Baptist Church life in several ways: the solution adopted in disciplinary cases when the nature of the offence was such that two witnesses could not be found (and when those witnesses were women), the inquisitorial proceedings of Churches that functioned as ecclesiastical tribunals whose legitimacy could be challenged by the accused, the uneasy relationships between London and the North, the emotional ties that were maintained despite ministerial wanderings, the role of the deacons as keepers of the manuscripts. Historians have persistently drawn a discreet veil over the Cripplegate events, and the way they were told in letters. Either the excommunication of David Crosley was considered too embarrassing to be retold, too specific to be of any interest, or too unclear to deserve comments. This has prevented commentators not only from tapping the rich vein of the Cripplegate records but also from realising that among post-Toleration Baptist ministers the figure of Crosley looms even larger than expected. Far from being a 'practical' Antinomian whose contempt of the moral law brought shame to a community (Whitley 1913, 109; MacDonald 1982, 119; Toon 1967, 152), Crosley was a gifted writer and a daring minister, questioning and challenging many aspects of Church discipline without straying from mainstream Calvinism. He was not once accused of doctrinal errings even by his worst enemies. In the course of a reassement of Crosley's career, it is worth paying attention to Hester Hannis's confession, to Susan Emerton's expressions of disgust, to Crosley's blatant lies, to Cripplegate's 'barbarity' and to Tottlebank's abuses. This means going back to Wheeler Robinson's 'anthropology' and to the Baptist Church records, not only for the evidence they yield about Church life but also for what they reveal about early eighteenth-century manuscript culture: their carefully worded narratives, their writing of history, their legal rhetoric balanced with the emotional charge of their epistolary exchanges.

References

Blomfield, W.L. 1912. Yorkshire Baptist churches in the 17th and 18th centuries. In *The Baptists of Yorkshire*, ed. C.E. Shipley, 53–112. Bradford/London: William Byles & Sons/Kingsgate Press.

Cooper Jr., James F. 1999. *Tenacious of their liberties: The congregationalists in colonial Massachusetts*. Oxford/New York: Oxford University Press.

Davies, Michael. 2009. Spirit in the letters: John Bunyan's congregational epistles. *Seventeenth Century* 25(2): 323–60.

Greaves, Richard L. 1975. The organizational response of nonconformity to repression and indulgence: The case of Bedfordshire. *Church History* 44: 472–84.

Halcomb, Joel. 2010. A social history of congregational religious practice during the Puritan revolution. Unpublished PhD thesis, University of Cambridge.

Hargreaves, James. 1816. *The life and memoir of the late Reverend John Hirst, forty two years pastor of the Baptist Church at Bacup: Also an appendix*. Rochdale: Joseph Littlewood.

Hayden, Roger. 2005. *English Baptist history and heritage*, published for The Baptist Union of Great Britain. Chipping Norton: Nigel Lynn.

Lewis, Evan R. 1893. *History of the Bethesba Baptist Church, Bardnoldswick, Yorks*. Cwmavon: L.L Griffiths.

MacDonald, Murdina D. 1982. London Calvinistic Baptists 1689–1729: Tensions within a community under toleration. Unpublished D.Phil thesis, University of Oxford.

Overend, Frederick. 1912. *History of the Ebenezer Baptist Church, Bacup*. London: Kingsgate Press.

Renihan, James M. 2008. *Edification and beauty: The practical ecclesiology of the English particular Baptists, 1675–1705*. Milton Keynes: Paternoster Press.

Toon, Peter. 1967. *The emergence of hyper-calvinism in English nonconformity, 1689–1765*. London: Olive Tree.

Wheeler Robinson, H. 1922–1923. Baptist church discipline 1689–1699. *Baptist Quarterly* 1: 112–28, 179–85.

Wheeler Robinson, H. 1924–1925. The value of denominational history. *Baptist Quarterly* 2: 100–112.

Whitley, W.T. 1913. *The Baptists of North-West England, 1649–1913*. London/Preston: Kingsgate Press/Toulmin & Sons.

Part II
Representations of British Catholicism

Chapter 6
'For the Greater Glory': Irish Jesuit Letters and the Irish Counter-Reformation, 1598–1626

David Finnegan

Historians have greatly exaggerated the importance of the Jesuits to the Irish Counter-Reformation. J.P. Mahaffy, a leading representative of this school of thought, argued that although there were never very many Jesuits in Ireland, those present were generals or organizers who exercised authority over the secular clergy and helped to turn the quarrel with England into one of religion rather than of race. In the process they 'effected the reconquest of a well-nigh lost province of the Roman Church' between 1560 and 1603. According to this interpretation by 1588—and referring to a period it is important to remember when there were no Jesuits in Ireland—the Society had so influenced the population that most in Ireland regarded the failure of the Armada as a disaster (Mahaffy 1903, xiii, 12–33). Mahaffy's misconceptions—shared by Richard Bagwell—were partially based upon over-familiarity with the state papers to the exclusion of almost all other documentary evidence. English government officials dealing with almost total religious recalcitrance in early modern Ireland were inclined to see Jesuits everywhere and regularly reported 'swarms' of them, when in fact there were few or none.[1] It was also predicated upon Irish Jesuit correspondence, which had been recently published—much in English translation—by Edmund Hogan, S.J. These are typified by the bombastic claim of 'Fr. Ibernus (Nicholas Leynich)' who wrote in 1607 that

> It would take me too long to describe the reformation and amendment of life and customs which have taken place since we made our entrance into this island; the oaths and blasphemies we did away with, the discords, enmities, and public adulteries, to which a remedy was applied through the public penance of the delinquents (Hogan and Fitzsimon 1881, 119).

[1] For some examples see TNA, SPI 63/154/37, 63/161/44.

D. Finnegan (✉)
Department of History, Warwick School, Warwick, CV34 6PP, UK
e-mail: dvdfnngn@gmail.com

A. Dunan-Page and C. Prunier (eds.), *Debating the Faith: Religion and Letter Writing in Great Britain, 1550-1800*, International Archives of the History of Ideas 209,
DOI 10.1007/978-94-007-5216-0_6, © Springer Science+Business Media Dordrecht 2013

While historians have retreated somewhat from Mahaffy's position, one leading scholar of the Irish Counter-Reformation recently wrote that 'considering that there were only four Jesuits actively on mission in Ireland in 1605, and that none of these were based in either Connaught or Ulster, it is obvious that however deficient the Old English [descendants of the medieval settlers from Britain] community was in clerical administration, the Gaelic regions must have been in a far worse condition' (Forrestal 1998, 31–2). This division of Irish Catholics into two groups also emphasized the differences between the development of reformed Catholicism in the native and old colonial sectors of Irish society; the Gaelic Church remaining hidebound as its 'Old English' counterpart took the impress of the Tridentine Reformation (Kearney 1960; Clarke 1978; Jackson 2007). This opinion has retained almost canonical status amongst historians of early modern Catholicism in Ireland. Yet such claims of the transformative influence of the Society of Jesus on early modern Irish Catholicism seem particularly outlandish when one looks at the reality of the earliest Jesuit efforts in Ireland.

Ireland was the Jesuit order's '*Missio primogenita*'. Alphonsus Salmeron and Paschase Broet arrived in Ireland on 23 February 1542. Although Francis Xavier had left Lisbon for India on 7 April 1541, he did not reach Goa until 6 May 1542 because of a lengthy stay in Mozambique. By this date the Jesuit mission to Ireland had already been abandoned. Its collapse was not due to poor planning. Ignatius Loyola had considerable input both in planning the mission and selecting the men who went to Ireland. Both were senior figures within the Society and within the Church. Salmeron was perhaps the leading papal theologian at Trent and Broet was instrumental in establishing the Society in France. Their disappointing failure in their 34 days in Ulster was shaped by the fact that the Irish lords' Catholic-coloured revolt had been quelled shortly before their arrival, and the lords had surrendered to Henry VIII's Irish agent, Sir Anthony St Leger. Fearing for their safety, Salmeron and Broet refused to meet with the Ulster chieftains and soon withdrew complaining of the province's uncivil and irreligious disposition. They described Ulster as a 'wild region', where they were often 'hungry and thirsty' and complained about the moral degeneracy of the Irish especially the numerous illegitimate relationships. But the reality is that the Jesuit missionaries were deeply shocked by their immersion in an alien culture, for which they lacked the language skills. They exhibited none of the understanding of local culture that later became the hallmark of the Jesuit organisation and were eager to leave; requiring a convincing reason for their withdrawal, they generalized negatively about Gaelic moral and political attitudes. In the words of William Bangert their report lacked 'completeness, balance and integrality' (Broderick 1940, 110–11; Bradshaw 1979, 54–6, 247–8; Bangert 1985, 167–71; McCoog 1996, 18–23).

This experience dampened the initial enthusiasm that Loyola had shown for the Irish project and was probably behind the lack of Jesuit interest in Ireland in succeeding decades. Nonetheless the experience had also been instructive. It made clear that any future success of the Jesuits as a missionary order in Ireland depended on developing a corps of Irish members of the Society (Mahaffy 1903, 25; McCoog 1996, 22–23). This was not a priority during the remainder of Loyola's Generalate and this ambivalence increased as the Society expanded rapidly after his death. And so Jesuit efforts in Ireland until the 1590s involved little more than the individual initiative of a handful of men; when David Wolfe was dispatched to Ireland in 1560 it was at the behest of Pope Pius IV, and his disgrace hardly boosted the Irish mission's profile (Morrissey 2004). This explains their limited impact in pastoral terms and the state's increasing suspicion of foreign clergy explains their involvement in the religio-political struggles of the late 1560s and 1570s (Morrissey 2004). The defeat of James Fitzmaurice Fitzgerald's rebellions and the execution of Edmund Daniel in 1572—the first Jesuit martyred in Europe—convinced Generals Lainez, Borgia, Mercurian and Aquaviva—at least in the early years of his Generalate—that the Jesuits would achieve little more than martyrdom in such a hostile environment. Thus it was not until 1598 that the Society attempted another missionary effort in Ireland. By now the Order itself had been transformed into a far larger, more structured and hierarchical institution, shaped by unswerving obedience to its General—at least in theory (O'Malley 1993; Martin 1973). This 'way of proceeding' also emerged as successive Generals sought to suppress those disruptive national impulses that William Allen observed were 'the common inclination of Adam' and distracted from the order's core mission (Knox 1882, 74; McCoog 2004).

Such national inclinations could not be suppressed entirely however and the Irish Jesuits appear to have agitated for a mission throughout the 1590s, possibly through fear of the new Protestant University founded at Dublin. They sought to take advantage of the alienation amongst their countrymen from the English government, then also distracted by the religiously coloured rebellion of the lords of Ulster (1594–1603) (Finnegan 2007, 70–110). The Ulster insurgents and their supporters in exile on the continent also appealed for an Irish mission and promised 'every protection and help, and certain lands for their maintenance' (Hogan 1884, 325). Eventually Aquaviva was persuaded, probably through the offices of Christopher Holywood, S.J, whom he described as 'a learned and mature man' and who was eventually appointed as mission Superior (Hogan 1884, 206; O'Donoghue 1981, 24). His instructions make it clear that this was an exclusively pastoral and apolitical mission. The General regarded Ireland as a 'vineyard' without (priestly) 'labourers', a situation that he had a duty to rectify (Hogan 1884, 176, 323). But in the mission's earliest years it was not entirely certain how that would function in practice: for although Aquaviva had made it clear that 'since the Irish Mission now existed it is a work for Irishmen', he clearly considered subordinating the Irish mission to the English vice-province (O'Donoghue 1981, 21).

In that light, Holywood's arrest in England *en route* to Ireland could have been a disaster for the fledgling mission. Aquaviva decided to appoint Richard Field, who had arrived in Dublin before Holywood's dispatch, Superior in Holywood's stead. In difficult circumstances Field did just about enough to keep the mission functioning. He wrote regularly to Aquaviva requesting more Jesuits for the Dublin area because 'this more civilized part of the kingdom requires work and the fruit will not be lacking' (Hogan 1880, 67–68). The reports of Archer and Fitzsimons suggesting the amazing possibilities for the order in Ireland also helped to sustain Aquaviva's enthusiasm for the mission (Hogan 1880, 38–40, 40–43). Progress was slow for Aquaviva had to balance the mission's needs against those of the wider society (Hogan 1884, 3–4). There is also the sense that he lacked the same confidence in Field as he reposed in Holywood—a problem not helped by the fact that during his period as Superior, Field never received any communication from his General (Hogan 1884, 169, 172–3, 177, 259), nor by his seeming inability to control Archer or Fitzsimons.

The appointment of Ludovico Mansoni, S.J., former Provincial of Naples, as papal legate to Ireland in May 1601 saw Aquaviva give more attention to the mission as this was an obvious opportunity to put the infant mission on a firm foundation (Jones 1953, 1–68). Mansoni was to be Superior upon his arrival in Ireland and three more Jesuits were dispatched to assist Field, who, after Archer's departure for Rome and Fitzsimons' arrest, was the only member of the Society then in Ireland (Hogan 1880, 79–81). In the event Mansoni never reached Ireland, and so, the mission limped along until the rebels' finally surrendered and Elizabeth died in the last days of March 1603. These events had two very important consequences; first Holywood was released and sent into continental exile from which he returned to Ireland in early 1604; and second the rebels' defeat made regular communication between the mission and the Generalate considerably easier for a time (Hogan and Fitzsimon 1881, 128).

Holywood's arrival in Ireland in March 1604 transformed the history of the Irish mission. Before his return the mission had largely been an individual operation with the Jesuits tending to work alone and from the homes of wealthy patrons.[2] This tended to reduce their scope for independent action especially if a patron had to be disciplined. These homes were usually in urban areas and this also tended to limit the scope of the mission to the towns. In a sense the Jesuits were becoming, as in England, a 'gentry clergy' (Haigh 1987). The lack of a proper mission structure before 1604 also allowed some of the more politically minded brothers to dabble in secular matters. This was something neither Holywood nor Aquaviva were willing to countenance and after Archer was conveniently dispatched to the continent in 1602 and Fitzsimons was expelled by the state in 1604, neither was allowed back. Holywood was keen to get the brothers to work in pairs once conditions were more settled and by 1606 Andrew Wise and Walter Wale were working from a residence (Hogan 1880, 190) and by the end of 1611 there was at least one residence in each province, a sure sign that the mission was moving towards permanence and autonomy.[3] Holywood also sought to promote men more rapidly within Ireland given that 'as most parts of Ireland are altogether Catholic … in 2 years they could know the Institute thoroughly' and as there was 'no hope of getting such men from the houses or colleges abroad' (Hogan and Fitzsimon 1881, 141).

This was part of Holywood's scheme to pressurize his General into putting greater resources at his disposal and possibly to have the Irish mission elevated into a fully-fledged province. His key problem was that, given the clandestine nature of Irish Catholicism, the greater popularity of the Franciscans and the Dominicans, and Rome's reluctance to sanction the transfer of religious properties that had formerly

[2] Royal Irish Academy, Dublin, MSS 23. F. 1, fols 305–06.

[3] Irish Jesuit Archive, Dublin, 5/D/7.

belonged to other orders, he was severely under-resourced at home. Above all else an organizer and communicator, Holywood was quick to identify the criticality of communication with the continent and especially with the General to advancing his goal. This persuaded the Irish Jesuits to embark on an intensive propaganda campaign. The aim was to show—as Fr. O'Kearney framed it in November 1605—that a 'great harvest is ripe, if we had reapers enough' (Hogan and Fitzsimon 1881, 124). They had to balance making clear the opportunities that existed given the support of the Irish people with the mission's dependence on the charity of the order's European provinces for training and then releasing Irishmen (Hogan and Fitzsimon 1881, 117, 120). Holywood's letters to the General, in which he claimed that 'it was generally affirmed, even by government officials, that we were the best reformers of the people', clearly suggested that more Jesuits in Ireland would mean further gains to the Catholic fold (Hogan and Fitzsimon 1881, 114, 119, 122, 123). Holywood made much of the fact that almost as many men (27) joined the order in the period 1596–1605, than in the previous 60 years. This increase gave the hope—expressed in almost every letter the brothers sent from Ireland—of a continued supply of Irish Jesuits with which to stock the mission (Hogan and Fitzsimon 1881, 48, 68–79, 118–19, 121, 124, 142; Hogan 1884, 170, 175, 209, 264–7, 324–5).

The Irish Jesuits then clearly hoped to take advantage of the rapid expansion of the Society. During Aquaviva's 34-year term (1581–1615) the number of Jesuits increased from 5,000 to 13,000; the number of schools increased from 144 to 372; residences from 33 to 123; provinces from 21 to 32. Expansion on this global scale posed new problems and increased the need for efficient organization. Consequently the society became more centralised and bureaucratised. The result of this institutionalization was a loss of direct contact with personal superiors and a consequent greater insistence on rules and procedures, a good number affecting communication via letter (Martin 1988, 107–11). The society had a well-established system of instructions for the writing of letters; the *Formula scribendi*, contained in the *Regulae Societatis Iesu* published under the generalship of Diego Laynez in 1561.[4] Thirty-five items detail when, why, and to whom letters should be written within the society (134–42). For example, superiors of residences and rectors of colleges were required to write a letter each week to their provincial leader. In this letter they were to catalogue the status of all those under their leadership both good and bad, their achievements as well as their setbacks so that the provincial might see 'everything as if he were present'. They were also required to write to their generals once every three months. Provincials in turn were to write to the superiors, rectors, and teachers of novitiates as well as to their generals every month (134–5). Some exceptions were made for distance or difficulty; provincials in India were to write whenever navigational opportunity presented itself, and rectors, superiors, and teachers in Brazil and New Spain were to write twice a year (135). Given the periodic bouts of repression that beset Ireland's Catholic community the Irish mission was also afforded some flexibility in this regard.

---

[4] I have used the 1604 edition in this essay.

The superiors of Jesuit residences and rectors of Jesuit colleges were required by the *Societiatis Iesu Regulae* to submit an annual report to their Provincial Superiors detailing the achievements of those under their control (138–40). The letter for a particular year was generally composed in the early weeks of the subsequent year. This took the form of a detailed report of the brothers' activities during the previous year and followed a relatively standard formula that to an extent dictated the nature and presentation of the information presented. It began with a general statement on the condition of the country, including any unusual episodes or events. Jesuits from around the province were to submit reports of their proceedings or as they did in early 1607 they might, if circumstances permitted, 'all come together in order to transmit ... the materials for the history of the year 1606' to their Superior who incorporated the information into a single report (Hogan and Fitzsimon 1881, 153). In the second catalogue the talents and qualities of each member of the province was described namely, innate disposition (*ingenium*), judgment (*iudicium*), wisdom (*prudentia*), experience in matters (*experientia rerum*) progress in letters (*profectus in litteris*), natural grasp of things (*naturalis complexio*) and to what ministry of the Society they were most suited (141). The logic was that the Superiors would have an abundance of information with which to cross-reference the reports. This would ensure not only the speedy resolution of disputes but reduce the danger of inaccurate or exaggerated information (Boswell 2003, 252).

These letters have long been an important source for historians studying the Society's history but recently their reliability has been questioned. A. Lynn Martin has revealed the inconsistencies, and even deceptions of Jesuit correspondence, asserting that they were capable of 'rhetorical exaggeration' in their writing (Martin 1988, 3–4). And although he and Grant Boswell have suggested that this point of view may reflect anachronistic pre-suppositions by failing to acknowledge the Jesuits' understanding of their own epistolary practice, he too accepts that their letters cannot always be taken at face value (Boswell 2003, 249). For Boswell the Society's expansion rested to a very considerable degree on its members' rhetorical strategies, especially their adaptive flexibility in the writing of letters. This entailed rejection of the slavish imitation of Ciceronian precepts and the employment of humanist motifs. Above all else this valued pragmatism. This ensured that Jesuit letters were often shaped to the rhetorical task at hand. This is typified in the eighteenth chapter of Antonio Possevino's 1593 work, *Bibliotheca selecta*, a compendium of Jesuit knowledge (and a precursor of the *ratio studiorum*).[5] Holding that 'There is, of course, one method of writing letters for republics, another for princes or monarchs and yet another for those who guide the Church for there is no single method of ruling people and handling affairs', Possevino sought to teach Jesuit novices to think about what they are writing, to whom, and to what end (Boswell 2003, 254). It appears that the members of the Irish mission incorporated this flexibility in their letters to their superiors on the continent in their pursuit of the support necessary to elevate their mission into a province. This is not to say that they departed too greatly from

[5] Antonii Possevini Societatis Iesu *Bibliotheca selecta de ratione studiorum in Historia, in Disciplinis, in salute omnium procuranda* (Rome, 1593).

Possevino's advice that the words and matter must be equal to the demands of reason and nature (Boswell 2003, 257). Perhaps those superiors were more aware than some historians have been of the rhetorical strategies in these epistolary writings.

Holywood's Annual letters were the chief vehicle through which his propaganda campaign was conducted. Holywood instructed his subordinates not to write to Rome without permission. This allowed a uniformity of style and purpose to be achieved as well as strengthening the Superior's authority (Jackson 2007, 199–203). He himself kept in regular contact with the brethren by letter and made sure Aquaviva was aware of this control. Holywood quickly established a network of correspondents to assist his project. As Jesuits were still not permitted in France he employed a secular priest to live in Bordeaux—a key port in Ireland's trade with the Continent—to manage his European correspondence though this was also occasionally sent through Spain and England (Hogan 1880, 134–5; O'Donoghue 1981, 64, 103). To protect the order and its business, Holywood sent a cipher to Rome in 1604 but he rarely used it and seemed to prefer allegory, aliases and obscure terms (Hogan and Fitzsimon 1881, 167). Direct connections with the provinces within the *monarquía* were also developed. These were essential for recruitment and training as most Irish Jesuits were professed in Spanish provinces. Holywood and others had regular correspondence with the secular priest, Christopher Cusack, who controlled the Irish colleges in the Spanish Netherlands and used his good offices to place novices in the Belgian novitiate (Hogan and Fitzsimon 1881, 68–79). Letters were also vital to persuading Irish Jesuits, then increasing in number across Europe, to work in 'our Japan' as well as seeking to persuade those provinces to which the men working in Ireland belonged of the merits of supporting the Irish mission (Hogan and Fitzsimon 1881, 140–41, 142, 178).

The nature of the correspondence, and more significantly the increasing unwillingness to send it through the English vice-province, reveals that the Irish mission feared they would be placed under an English Superior. In this period it is clear that both the Jesuit Curia and the Roman authorities more generally regarded Ireland as an adjunct to England. Until the 1620s most of the papers and correspondence on Irish affairs were deposited along with the much larger bulk of material concerning England in the Roman archives. The Irish missionaries—proud of the ecclesiastical history of their nation (which they completely identified as Ireland)—opposed subordination to another jurisdiction (Hogan 1880, 161–2). The English Jesuits were also less than enamoured by the idea given the involvement of some of their Irish *confrères* with the Irish rebels and by a certain cultural disdain towards the Irish in general (Jackson 2007, 195; Highley 2008).[6] But continental provincials and heads of educational institutions, especially those within the *monarquía* supported the absorption of the Irish mission by the English vice-province (Hogan 1884, 351–2). These were reluctant to use scarce funds to form Jesuits who were then likely to be called away to the Irish mission. Direct and frequent contact with the General was thus deemed critical to the increase of the mission: for keeping Aquaviva informed of local needs and opportunities increased the chances of being given greater resources and support.

[6] See also W. Camden, *Brittania* (1607), 788–93.

This correspondence did persuade Aquaviva that the prospects for the Jesuits in Ireland demanded his support and he informed the Provincials in 1604 that with regard to the Irish Jesuits in Europe the requirements of the Irish mission took precedence over those of the Provinces (Hogan 1884, 286–7). Yet in practice this support from the General was not always decisive. Some provincials dragged their feet interminably, others pleaded extenuating circumstances and others simply baldly refused to comply (Hogan 1884, 368). No amount of pressure could induce the Provincial of Valladolid to release William Bathe from his teaching duties for service on the mission, nor would the Austrian Province part with Florence O'More on the grounds that his presence in Neuhaus was critical to its temporal and spiritual well-being (Hogan 1884, 123–4, 382). Occasionally Irish frustration could lead to heated outbursts. Holywood even went as far as warning that on the day of judgment he would accuse George Duras, the German Assistant, 'of the loss of all the souls that would have been won over … in this country by the industry of Gerot', an Irish Jesuit Holywood had earmarked for the mission. Nor was Holywood above using moral pressure on his General: 'If your paternity wishes a copious harvest, see that nature helps grace, and that we may have a residence in the North' (Hogan and Fitzsimon 1881, 141, 166). Aquaviva's enthusiasm for the mission then threatened to be limited by the need to balance it with those of the Provinces. Indeed although the situation became ever more favourable for the mission after 1604, it was not until 1652, at the tenth General Congregation of the order, that it was resolved that every province should always have one Irish Jesuit in training at its own expense for the Irish mission (Hogan 1884, 3).

Such were the imperatives that shaped the nature of Irish Jesuit correspondence. This essay's central contention is that this correspondence, designed to suggest that the Jesuits were indispensable to the salvation of the Irish population, while reflecting the very valuable work that the Society engaged in during the period, has also cast them as the saviours of Irish Catholicism. Indeed it is a historiographical sacred cow that the other clergy merely served as keepers of the tender flame of a survivalist Catholicism until the Jesuits arrived with the manna of post-Tridentine Catholicism. There is a strong sense that the Irish Jesuits may well have bought into the order's mystique to an unhealthy extent. Fitzsimons described his *confrères* as the 'most behated by the precursors of Antichrist' and reveled in their reputation amongst the heretics as 'janissaries of the Pope'. He elevated his order above the other regular and secular clergy in Ireland when he pointed out that during his incarceration 'one bishop, three Franciscans and six secular priests recovered their liberty' through various means but because he 'was a Jesuit … that name so terrible and hateful to the enemies of God … no fair nor foul means could get freedom for [him]' (Hogan 1884, 204, 219). His brothers held a similarly elevated view of their Society's image and of their achievements with regard to securing adherence to Catholicism. Writing in April 1604, Fitzsimons eulogized the work of the Irish Jesuits and claimed that their effect on the Irish population was quite profound: 'those, who before were mere *tabula rasae*, know the teaching of the faith, and piety flourishes where all had been a waste' (Hogan 1884, 261).

When this correspondence is tested however it is often possible to see its claims contradicted by the letters of other Jesuits, members of other orders, the secular clergy or government agents. An excellent example was the 1604 letter of Nicholas Leynich and Andrew Mulroney, then working in Munster (in southwest Ireland), to their Portuguese *confrères*. Therein they described their work of reconciliation, which included the conversion of Richard Burke, third earl of Clanricarde and of Theobald Butler, the earl of Ormond's heir. They also refuted heretical propositions concerning the Real Presence, purgatory, invocation of the saints, papal supremacy, the Mass as a sacrifice and the Latin liturgy, using texts from Scripture and the Fathers. They also claimed to have encouraged restitutions, composed family quarrels, dissuaded people from attending Protestant Churches, reduced swearing and blasphemy and heard many thousands of confessions from penitents who had never previously confessed (O'Donoghue 1981, 36–7).

This letter offers an excellent example of the tendency of the Irish Jesuits to amplify their own achievements and disregard those of others. For instance of the two noblemen they claimed to have converted from Protestantism, Burke was in reality a longstanding Catholic, and Butler remained a committed Protestant until his death in 1613. Moreover, their claim that they rested secure in the support of the people, whose perseverance in Catholicism they regarded as extraordinary considering the treatment they had received from the government in recent decades, clashes with their assertion that the Irish laity lacked any knowledge of the rudiments of faith and were constantly consorting with heretical ministers. Neither do Jesuit claims of Irish Catholic irregularity always dovetail with their claims for their mission's successes. Working in north Munster in 1605, Leynich and Mulroney claimed that they had heard so many new penitents confess 'they could hardly breathe' (O'Donoghue 1981, 40). The year before however, Holywood had visited the area and reported that owing to the order's efforts the Catholics there were openly professing their faith and exhibiting a fairly satisfactory standard of instruction (Corboy 1941, 86).

The Jesuits certainly missed or more likely downplayed the deep attachment of the Irish to the Franciscan and (to a lesser extent) the Dominican friars. These, as historians have recently stressed, played a key role in frustrating the advance of Protestantism as they were 'numerous, widely dispersed, pastorally dynamic, respected by the laity of all social degrees, and well attuned to the vernacular as a mode of evangelisation'. More importantly these orders maintained a continual presence in Ireland during the era (Bradshaw 1998, 100–102). Even in Munster, where the Jesuits devoted much of their effort and limited resources, it was the friars whom the members of the Protestant Church blamed for the refusal of the population to attend the 'church by law established' (Hogan and Fitzsimon 1881, 166–7). Yet again however it appears that the historians of these other orders may also have unwittingly downplayed the enormous role of the secular clergy in securing Ireland for Catholicism (Bradshaw 1979, 208–09; Jefferies 1997). Dr. Dermot Creagh, titular bishop of Cork and Cloyne, having returned to Ireland in 1579 had spent the next 24 years swearing the people of Munster to the Pope, to have nothing to do with the 'heretics' and exercising 'all manner of spiritual jurisdictions in the whole

Province, being the Pope's legate, consecrating Churches, making Priests, confirming children, deciding matrimony causes'.[7] This had brought about a profound reawakening of popular Catholicism in the province with many Irish-born clerics who had served as ministers of the established Church apostatizing in the late 1580s and early 1590s and it was held to be critical in the alienation of the people of Munster from the state during the 1590s conflict (Ford 1985, 24–6; Finnegan 2007, 86–9).

Too literal a focus on the Irish Jesuits' correspondence has exercised a distorting effect on early modern Irish historiography and has led to a false dichotomy between the religious mores and observances of the Gaelic Irish and the 'Old English'. Essentially it is argued that the Jesuits brought a modern Tridentine mode of Catholic observance to the latter and that the Church in the Gaelic regions dominated by the older orders, remained hidebound and blindly dedicated to more archaic modes of moral observance. The effect was to widen the gap between the two communities. This is a complete misnomer. In point of fact the Church in the Gaelic regions was well in advance of the 'Old English' areas in terms of implementing a Tridentine programme. While the first synod to implement Trent's decrees in what might be called an 'Old English' region did not meet until 1614, the decrees were promulgated in Ulster and Connacht in the late 1560s (Hogan and Fitzsimon 1881, 143; Finnegan 2007, 18). On the basis of little evidence, it has been suggested that any such Tridentine emphasis would have been impossible, as the Gaelic regions were not adequately provided with a parish clergy. Yet given that the government had not extended its power in anything other than a nominal sense in large swathes of Ireland until well into the seventeenth century, it seems reasonable to suppose that the necessary parish structures for the bishops to impose their will upon clergy and laity remained intact (Jefferies, 1997 182–3; Finnegan 2007, 71–8). Especially given that the clergy were 'almost honoured as gods'.[8] Even after the severe attenuation of the personnel and fabric of the Church in Ulster caused by the Nine Years' War there was a resident priest in almost every parish in the diocese of Derry (Jefferies 1999, 175–204).

Although the Jesuit Annual letter of 1611 suggested that the people living on the Connacht-Ulster border in northwest Ireland had not received the sacraments in living memory this clashes with the recollection of the Armada survivor Captain de Cuellar during his perambulations in this exact area a generation earlier. He noted that mass amongst the Irish was 'regulated according to the orders of the Church of Rome' and was conducted in a familiarly orthodox manner (Allingham 1897, 32–4). It also ignores the fact that Thomas Mulkerrin and two other Jesuits had extensively ministered in the area the year previously (O'Donoghue 1981, 81–3). For despite adverse political circumstances which rendered the consistent implementation of post-Tridentine Church law difficult, the Irish hierarchy and clergy were anxious to observe as best they could correct canonical form and procedure in their ministrations throughout the Tudor period—with the support and

---

[7] TNA, SPI 63/202/iii/144, 63/207/i/108; Bodleian Library, Oxford, Rawlinson MSS, C. 98, fols 26v–27r.

[8] TNA, SPI 63/169/20.

understanding of Rome—an arrangement continued into the seventeenth century (Canonical faculties). The activities of Ulster's episcopacy were indicated by the following report on the activities of Redmund O'Gallagher, Bishop of Derry. He, being

> legate to the Pope and Custos Armaghnen ... [and] one of the three Irish Bishops that were in the Council of Trent ... used all manner of spiritual jurisdiction throughout all Ulster, consecrating Churches, ordaining Priests, confirming children, and giving all manner of dispensations, riding with pomp and company from place to place as it was accustomed in Queen Mary's days.

His ministry greatly influenced 'all the rest of the Clergy there'.[9] In summer 1593, William Lyon, the zealous Protestant Bishop of Cork, Cloyne and Ross (1583–1617) in the very south of the island, dolefully reported to Lord Deputy William Fitzwilliam that Bishop Eugene O'Hart was conducting confirmation services throughout the country and that people flocked to him to have children confirmed and baptized (Moran 1865, 214).[10] This is not to suggest that Gaelic Ireland was a model Tridentine heartland *à la* Milan, but rather to remember that the process of reform across Europe was a slow one that often utilized existing personnel and institutions (Forster 1992; Poska 1998; Ryan 2003). We would do well to remember John Bossy's cautionary caveat that post-Trent Catholic reform 'did not require much new legislation but called for a decidedly new attitude to old legislation' (Bossy 1970, 53). Even more importantly it reveals that the Irish population were hardly as starved of clergymen as the Jesuit missioners claimed in the first decade of their mission.

The Annual Letters often stressed that forward thinking parochial clergy assisted the Jesuits and that they were frequently asked to go on diocesan visitations. They also claimed that many secular priests took a step forward to a distinctively Jesuit form of spirituality by making the Spiritual Exercises (Hogan 1880, *passim*). This may be true but beyond the Archbishop of Cashel, David O'Kearney (brother of Barnaby O'Kearney, S.J.), and later David Rothe, Vicar Apostolic of Armagh and Thomas Dease, Bishop of Meath the order had few friends amongst the secular clergy because they competed with them for precious resources from the Catholic laity (Hogan 1884, 171–2, 1880, 164–6). Aquaviva constantly advised prudence and modesty in their relations with the secular clergy. He advised that the Jesuits take mass offerings to avoid giving off an offensive odour of sanctity and offending secular clergy, provided they give those gifts to the poor as soon as possible (O'Donoghue 1981, 94). Nor did the Irish Jesuits directly acknowledge that much of their revenue was derived from alms or mass offerings as it was forbidden for Jesuits to receive payment for the performance of religious duties (Jackson 2007, 213–14).

Although they wrote at length about the need for new men to gain converts at all levels of society they were still able to leave two of their number with the earl of Ormond for prolonged periods to ensure his conversion (Hogan 1884, 333–4).

---

[9] Bodleian Library, Oxford, Rawlinson MSS, C. 98, fols 26v–27r.

[10] TNA, SPI 63/183/161.

This was of a piece with their order's longstanding focus on proselytizing elites, but it also somewhat contradicts their claim to have played a key role in re-establishing popular Catholicism. Indeed a recent study of the Irish Jesuits has concluded that the mission was one closely associated with the 'better sort' and whose missioners had scant interest in the 'mere Irish' (Jackson 2007, 235–43; see also O'Donoghue 1981, 62–88). The missioners' own observations attest to the longstanding attachment of the Irish to Catholicism (Hogan and Fitzsimon 1881, 141, 145, 155). In 1611, in a letter to Christopher Cusack, Fitzsimons observed that even when heresy was first reintroduced to Ireland early in Elizabeth's reign 'it was by force, never by voluntary allowance' (Hogan and Fitzsimon 1881, 68). Indeed in 1609, Archbishop O'Kearney wrote that 'scarcely one in a thousand of the Irish Nation is tainted with heresy' and 'of all the barons, viscounts and earls of Irish birth, hardly three can be found who, for any consideration, have abandoned the faith'. Even those 'political dissemblers' who have embraced Protestantism have done so for political reasons and 'send for the priest in the evening of life' (Hogan and Fitzsimon 1881, 194).

Perhaps the subtlest Jesuit insinuation was that they were responsible for steeling the populations of the towns to resist the government's determined effort to enforce religious conformity between 1605 and 1607. Before this bout of persecution commenced, Holywood claimed he had assembled Dublin's aldermen to explain the issues at stake in the forthcoming struggle. Urging them to withstand the 'fury of the wolves' he explained the 'nature and object of the persecution' and warned that 'we are bound to profess our faith before kings and princes and the magistrates of the earth'. Throughout the following months, he remained vigilant responding to queries sent from the prisoners, dispelling doubts and praising their 'constancy' (Hogan and Fitzsimon 1881, 142, 149, 150, 152, 154, 156, 1894, 179–80, 182). While he was certainly involved in galvanizing recusant resistance it seems that other clergy played their part. In a later letter he admitted however that the Dubliners would not permit his presence lest he be arrested or further inflame their persecutors (Hogan 1884, 439). Fr. O'Kearney's letter to Peter Lombard, Archbishop of Armagh, of 4 October 1606 also went a little off script: he described the constancy of the recusants of Carrick-on-Suir and Waterford in the face of government pressure but did not claim that this owed to his presence. Indeed he was forced to flee both towns lest he and his *confrères* be taken. It was the secular priests that remained to give solace to the beleaguered inhabitants (Hogan and Fitzsimon 1881, 144–6).

This testifies to the fact that the battle for the hearts and minds of the critical mass of the Irish people had been decided in Rome's favour long before the Jesuit mission was instituted. Indeed as early as 1570 Sir Henry Sidney noted that religion was a key aspect of the failing relationship between the state and the Irish towns.[11] In 1577 Sir William Drury complained that the inhabitants of Waterford maintained a 'great numbers of students of this city' at Louvain and were 'so cankered in Popery, undutiful to her Majesty and slandering the Gospel publicly' that they celebrated 'Masses

[11] TNA, SPI 63/30/56.

infinite in their several churches' (Brady 1868, 22–24). By 1600 government servitors contended that such was the 'obstinacy, boldness, and contempt' of the papists that 'were it not that the Deputy keepeth the state in these parts about Dublin, and is attended with forces, as is meet, we see just cause to fear that massing and idolatry would be brought into, and erected even in the cathedral churches, within these civil cities and towns...'[12] Henry Fitzsimons claimed in 1604 that 'religion does not strike deep and firm roots here; people, by a kind of general propensity, follow more the name than the reality of the Catholic faith, and thus are borne to and fro by the winds of edicts and threats'. This jars with the fact that twenty years earlier the Old English community had comprehensively rejected in parliament an effort to enact penal legislation in Ireland and had openly committed to recusancy (Hogan and Fitzsimon 1881, 59; Treadwell 1985; Brady 1985). Here Fitzsimons was denying his own background and that of his *confrères* given that had their parents not made the conscious choice to send their children to the continent for a Catholic education rather than to England for a Protestant one, Catholicism amongst the 'Old English' would have withered (Hammerstein 1971; Hogan and Fitzsimon 1881, 68–69). It is evidence of a continuing distaste for the Reformed religion inculcated by the survivalist Church in Ireland in the second half of the sixteenth century rather than of a sudden rejection inspired by the Jesuits in the early seventeenth.

The correspondence of the Irish Mission give the impression of a hard working and well led group going 'hither and thither according to our instructions' doing 'no small amount of good' in their 'no less glorious than fatiguing mission' (Hogan and Fitzsimon 1881, 118, 176, 178). Yet as this essay has argued they must be read cautiously and against a wider range of sources because they were clearly influenced by an *arrière-pensée*; to demonstrate that its missioners should be supported by the European provinces and maintained independently of the English vice-province. Thus they are in essence a litany of success with any difficulties or failures mentioned only to embellish the eventual triumph. Such failures as most certainly occurred had no place in the reports of a mission fighting for its very existence. They are also problematic because of the deliberate obfuscation of the names of those individuals they interacted with and the places and dates of events to protect identities lest the letters fall into the hands of the state. Such caution was perhaps necessary, but it frustrated contemporaries—including Aquaviva—as much as it has confounded historians. The lack of detail also allowed the Irish Jesuits to present the information that they wanted to without fear of contradiction and it made it very difficult to dispute their claims that they were the order most capable of introducing Tridentine reform to the Irish masses. This gave them a better claim to the finite resources of the Society.

Their apparent success in difficult circumstances and the promise of further gains if more support was forthcoming made Aquaviva increasingly happy to facilitate the mission. He endeavoured to make available regular places for the Irish Jesuits in the European provinces and to free Irishmen in those provinces for service at home.

[12] TNA, SPI 63/207/ii/92.

In 1609 there was a major reorganization of the mission's structure and significantly the instructions were the same as those given the English Superior (O'Donoghue 1981, 40, 95–7). Thomas White was appointed procurator with responsibility for fundraising for the mission on the continent while Holywood was asked to nominate suitable men for the posts of socius, consultor and admonitor in Ireland. In 1610 Holywood sent Walter Wale to Rome to present a plan for the growth of the mission. It was obviously felt that progress was too slow and that the European provincials were still far from fully cooperating with the mission. By 1611 the Superior's request for agents at Rome, Bordeaux and Douai had been granted; appointments which lasted into the 1620s when the Irish hierarchy was reconstructed (O'Donoghue 1981, 97–9). The mission was also prospering with increasing support coming from wealthy patrons and could afford to be far more discerning with regard to those they professed as increasing numbers sought to join their ranks (Jackson 2007, 190–91). For all that progress in numerical terms was painstakingly slow. In 1610 there were 18 brothers in Ireland, 10 of whom worked as priests out of a total of 72 Irish members of the order (Hogan 1880, 228). And although there was a spurt of growth in the years immediately after Wale's Roman visit by 1623 there were still only 40 Jesuits in Ireland compared to 800 secular priests, 200 Franciscans and 20 Dominicans (Martin 1962, 245).

The question as to how the Irish shifted from being church papists in the 1560s to convinced recusants by the 1580s (at the latest) remains somewhat unclear, but what is apparent is that the Society of Jesus had little active part to play in the transformation within Ireland. Few contemporaries saw the Irish Jesuits in the heroic role they cast for themselves. After the government clampdown on the Catholic mass-houses and schools in Dublin in 1630, John Roche, Bishop of Ferns and no friend to the Franciscans, noted that 'the Jesuits were not so forward as the friars in opening their schools or oratories; and you know they judge it prudence to suffer others try the ford before them' (Jennings 1953, 333). This essay has not sought to suggest that the Irish Jesuits did not do sterling missionary work in early modern Ireland. It is clear that they did. But it is also clear that through writing regularly, and not always truthfully, the Irish Jesuits greatly advanced their contemporary aims and in historical terms laid claim to a disproportionately large impact on the development of Catholicism in early modern Ireland. The editing of their letters by Edmund Hogan in the 1880s gave them a head-start in historiographical terms and allowed a mission numbering 51 in 1641 to claim a status rarely afforded to the Franciscans, an order with twenty times more members. This success owed much to the missioners' ability to present a united front and to trail the progress that a well-supported mission might make, but most to the willingness of Aquaviva to be swayed by the information presented. Little wonder then that Henry Fitzsimons thanked Aquaviva so profoundly in his 1611 treatise on the Mass: 'I proclaim that I am greatly indebted to you for the immense services rendered to myself and to my country … To us you have been not only a Father General … but you have wished to be our Father Assistant by the special care you have taken of us' (Hogan and Fitzsimon 1881, 81).

References

Allingham, Hugh. 1897. *Captain Cuellar's adventures in Connacht and Ulster, 1588 AD*, to which is added an Introduction and complete translation of Captain Cuellar's narrative of the Spanish Armada, ed. R. Crawford. London: Elliot Stock.

Bangert, William W. 1985. *Claude Jay and Alfonso Salmeron, two early Jesuits*. Chicago: Loyola University Press.

Bossy, John. 1970. The counter-reformation and the people of Catholic Europe. *Past & Present* 47: 51–70.

Boswell, Grant. 2003. Letter writing among the Jesuits: Antonio Possevino's advice in the 'Bibliotheca Selecta' (1593). *Huntington Library Quarterly* 66: 247–262.

Bradshaw, Brendan. 1979. *The Irish constitutional revolution of the sixteenth century*. Cambridge: Cambridge University Press.

Bradshaw, Brendan. 1998. The English reformation and identity formation in Ireland and Wales. In *British consciousness and identity: the making of Britain, 1533–1707*, ed. Brendan Bradshaw and Peter Roberts, 43–111. Cambridge: Cambridge University Press.

Brady, W.Maziere. 1868. *State papers concerning the Irish church in the time of Queen Elizabeth*. London: Longmans, Green, Reader & Dyer.

Brady, Ciaran. 1985. Conservative subversives: The community of the Pale and the Dublin administration, 1556–86. In *Radicals, rebels and establishments*, ed. Patrick J. Corish. *Historical Studies* 15: 11–32.

Broderick, James. 1940. *The origin of the Jesuits*. London: Longmans.

Clarke, Aidan. 1978. Colonial identity in early seventeenth-century Ireland. In *Nationality and the pursuit of national independence*, ed. T.W. Moody. *Historical Studies* 11: 57–71.

Corboy, J.J. 1941. The Jesuit mission to Ireland 1596–1626. Unpublished M.A. thesis, University College Dublin.

Finnegan, David. 2007. The impact of the counter-reformation on the political thinking of Irish Catholics, c.1540–c.1640. Unpublished PhD thesis, University of Cambridge.

Ford, Alan. 1985. *The protestant reformation in Ireland, 1590–1641*. Frankfurt-am-Main: Peter Lang.

Forrestal, Alison. 1998. *The Catholic Synods in Ireland, 1600–1690*. Dublin: Four Courts Press.

Forster, Marc. 1992. *The counter-reformation in the villages: Religion and reform in the Bishopric of Speyer, 1560–1720*. Ithaca/London: Cornell University Press.

Haigh, Christopher (ed.). 1987. *The English reformation revised*. Cambridge: Cambridge University Press.

Hammerstein-Robinson, Helga. 1971. Aspects of the continental education of Irish students in the reign of Elizabeth I. *Historical Studies* 8: 137–54.

Highley, Christopher. 2008. *Catholics writing the nation in early modern Britain and Ireland*. Oxford: Oxford University Press.

Hogan, Edmund, (ed.). 1880. *Ibernia Ignatiana seu Ibernorum Societatis Iesu patrum monumenta collecta secundum annorum ordinem distributa, notis illustrata, et scionum elogiis quota a P. Edmundo Hogan*. Dublin: Excudebat Societas Typographica Dubliniensis.

Hogan, Edmund, and H. Fitzsimon (eds.). 1881. *Words of comfort to persecuted Catholics. Written in exile, anno 1607 & letters from a cell in Dublin Castle and diary of the Bohemian war of 1620*. Dublin: M.H. Gill & sons.

Hogan, Edmund. 1884. *Distinguished Irishmen of the sixteenth century*. London: Burns and Oates Limited.

Jackson, B. 2007. The role and influence of the Irish Missions of the Society of Jesus on the implementation of a counter-reformation among the old English in Ireland, 1542–1633. Unpublished PhD thesis, Trinity College Dublin.

Jefferies, Henry A. 1997. *Priests and prelates of Armagh in the age of reformations, 1518–1558*. Dublin: Four Courts Press.

Jefferies, Henry A. 1999. Derry diocese on the eve of the plantation. In *Derry and Londonderry history and society*, ed. G. O'Brien, 175–204. Dublin: Geography Publications.
Jennings, Brendan (ed.). 1953. *Wadding papers, 1614–1638*. Dublin: Stationery Office.
Jones, F. M. (ed.). 1953. Correspondence of Fr. Ludovico Mansoni, S.J., papal nuncio to Ireland and Papal briefs to Fr. Mansoni, papal nuncio to Ireland. *Archivium Hibernicum* 17: 1–50, 51–68.
Kearney, Hugh. 1960. Ecclesiastical politics and the counter-reformation in Ireland, 1618–1648. *The Journal of Ecclesiastical History* 11: 202–212.
Knox, Thomas Francis (ed.). 1882. *Letters and memorials of William Cardinal Allen*. London: David Nutt.
Mahaffy, John Pentland. 1903. *An Epoch in Irish History: Trinity College Dublin—Its foundation and early fortunes, 1598–1651*. London: Unwin.
Martin, Francis X. 1962. *Friar Nugent: A study of Francis Lavalin Nugent (1569–1635), Agent of the counter-reformation*. Rome/London: Capuchin Historical Institure.
Martin, A. Lynn. 1973. The Jesuit Mystique. *Sixteenth Century Journal* 4: 31–40.
Martin, A. Lynn. 1988. *The Jesuit mind: The mentality of an Elite in early modern France*. Ithaca: Cornell University Press.
McCoog, Thomas M. 1996. *The society of Jesus of Ireland, Scotland, and England 1541–1588: 'Our way of proceeding'?* Leiden: Brill.
McCoog, Thomas M. (ed.). 2004. *The Mercurian project: Forming Jesuit Culture 1573–1580*. Rome/St Louis: Institutum Historicum Societatis Iesu; Institute of Jesuit Sources.
Moran, Patrick Francis. 1865. The see of Achonry in the sixteenth century. *Irish Ecclesiastical Record* 1: 209–217.
Morrissey, Thomas J, S.J. 2004. 'Almost hated and detested by all'? The problem of David Wolfe. In *The Mercurian project: Forming Jesuit Culture 1573–1580*, ed. Thomas M. McCoog, 675–704. Rome/st Louis: Institutum Historicum Societatis Iesu; Institute of Jesuit Sources.
O'Donoghue, Fergus M. 1981. The Jesuit mission in Ireland, 1598–1651. Unpublished PhD thesis, Catholic University of America.
O'Malley, John W. 1993. *The first Jesuits*. Cambridge, MA/London: Harvard University Press.
Poska, Allyson M. 1998. *Regulating the people: The Catholic reformation in seventeenth-century Spain*. Leiden: Brill.
Regulae Societatis Iesu. 1604. Lugduni [Lyon]: Jacobi Roussin.
Ryan, Salvador. 2003. Popular religion in Gaelic Ireland, 1445–1645. Unpublished PhD thesis, NUI, Maynooth.
Treadwell, Victor. 1985. Sir John Perrot and the Irish parliament of 1585–1586. *Proceedings of the Royal Irish Academy* 85, sect. C: 259–308.

Chapter 7
Negotiating Catholic Kingship for a Protestant People: 'Private' Letters, Royal Declarations and the Achievement of Religious Detente in the Jacobite Underground, 1702–1718

Daniel Szechi

Given that most of those subject to the princes of early modern Europe never saw their sovereigns in the flesh, but rather encountered them through their formal statements, declarations and images on coins, communications between the exiled Stuart dynasty and its putative subjects in the British Isles were not physically outwith the normal range of encounters between rulers and ruled. Few of them would ever see the king-in-exile, but there again few of the subjects of Peter the Great or Philip V of Spain ever saw them either. The Stuarts were, then, not exceptional in communicating with their people via letters, declarations and visual propaganda, or in receiving in return a stream of letters, memoranda and petitions from the Jacobites[1] of the three kingdoms.[2]

What was different about communications between the Jacobite kings and their subjects was that it took place in a hidden, parallel world mentally, and often physically, separated from the everyday business of life. For most Jacobites there was a monarch and a government whom you acknowledged in the courts and in public and to whom you paid taxes, and there was a king over the water who ruled your secret heart. And the contours of your relationship with that special, distant king in large part stemmed from understandings reached through the exchange of declarations, letters and memoranda smuggled in and out of the British Isles. The logical corollary of this 'literary' quality to the relationship between rulers and 'ruled' was that the exact text of the statements made by the exiled kings assumed a very special significance. Both sides exercised particular care draughting and presenting missives

[1] The term 'Jacobite', derived from *Jacobus* the Latin translation of James (as in James II and VII), referred to all the secret (and some not so secret) supporters of the exiled dynasty.

[2] See for examples of these: Historical Manuscripts Commission, *Calendar of the Stuart Papers Belonging to His Majesty the King Preserved at Windsor Castle*, 8 vols (1902–20; thereafter HMC, *Stuart*), i. 218–20, 236, 343–5, 361–2, 520–4.

D. Szechi (✉)
Department of History, University of Manchester, Oxford Road, Manchester M13 9PL, UK
e-mail: daniel.szechi@manchester.ac.uk

A. Dunan-Page and C. Prunier (eds.), *Debating the Faith: Religion and Letter Writing in Great Britain, 1550-1800*, International Archives of the History of Ideas 209,
DOI 10.1007/978-94-007-5216-0_7,

outlining their positions and searched each other's correspondence for nuances that easily escape the modern reader. The Stuart kings and their erstwhile subjects wanted to be very sure they understood each other.[3]

This was not the only unusual aspect of their relationship. The Stuarts in exile tried very hard to maintain the formal, ceremonial and visual attributes of contemporary royalty. They kept as large a court as they could afford and there lived in as appropriate a style as they could. They patronized artists and musicians, they distributed charity and patronage as far as their means allowed and they set an example to their subjects by publicly professing their faith through daily attendance at religious services (Corp et al. 2004, 76–256 *passim*). What separated them from their peers was a central truth that could never be officially acknowledged by James II and VII and his heirs: the Jacobite communities' allegiance to the Stuarts was voluntary. Being a Jacobite was a conscious choice which defied law and government in the British Isles. It was intrinsically dangerous to anyone who harboured, and even more so if they openly expressed, such an allegiance and there was little the Stuarts could do to help their followers who fell foul of the laws demanding untrammelled acceptance of the post-1688 state (Monod 1989, 233–66). The upshot of which was that the exiled Stuarts could only rarely command their putative subjects. Instead they needed to persuade them to heed their true kings. In truth, all contemporary European monarchs (even so-called 'Absolute' monarchs) needed to persuade, rather than force, their subjects most of the time (Bonney 1991, 486; Bohanan 1992, 109–14; Hubatsch 1975, 158–68). But for the exiled Stuarts the second option virtually did not exist. This profoundly influenced their relationship with their adherents in the British Isles, and nowhere more so than in their fraught exchanges over religion.

The fundamental religious problem for the Stuarts was that the exiled royal family was piously, devoutly Catholic up to 1750, when Charles Edward for a time converted to Anglicanism (McLynn 1988, 399). More typically, after 1692 James II and VII led a conspicuously holy life that became the basis of his candidature for canonisation in 1734 (Miller 1977, 240). By contrast, the vast majority of the Jacobites in England and Scotland were Protestant, and though there were a great many Irish Catholics who regarded the Stuarts as their rightful monarchs (and yearned, fought and died for them 1688–1760) they counted for little in the internal politics of the Jacobite movement. Strategically England was the glittering prize, and because Scotland could act as a platform for winning it the more militant Scots Jacobites enjoyed at least equal status with their English peers (Szechi 1994, 18–23, 2010, 43). It was, however, a commonplace in both kingdoms that they could not be prosperous and at peace if they were ruled by a monarch who was not a Protestant. Hence according to Whitelocke Bulstrode it was an 'absurdity' for any Protestant to

[3] For examples of which see: *Memoirs of Thomas, Earl of Ailesbury. Written by Himself*, ed. W. E. Buckley 2 vols (*Roxburghe Club*, 1890), i. 319, 325; *Letters of George Lockhart of Carnwath, 1698–1732*, ed. Daniel Szechi (*Scottish History Society*, Edinburgh, 1989), pp. 181–2: Lockhart to James III and VIII, [Dryden?] 7 Dec. 1722.

support James Stuart, the Old Pretender (the son of James II and VII and James III and VIII in the Jacobite line of succession).[4] For another pamphleteer it was self-evident that, 'the papists are endeavouring to divide us, in order to ruin us', and that, 'their religion obliges them to extirpate ours'.[5] In the event of a Jacobite victory, warned John England, 'We must all of us then turn papists, or be burnt at a stake, as in the days of Queen Mary, or else be put to other cruel torments, and inhumane treatment.'[6] The exiled dynasty also had a bad legacy to overcome: all Protestants in the British Isles, Jacobite and non-Jacobite, had bitter memories of James II and VII's catholicizing policies 1685–1688, and the threat these had posed to the Protestant communities of the British Isles was endlessly rehearsed in sermons, pamphlets and newspapers for the next century (Harris 2007, 182–236).

Overcoming such ingrained hostility and suspicion towards their religion in England and Scotland was always going to be very difficult for the exiled Stuarts, and it was little less of a problem among the Jacobites than among the population at large. Their difficulties were, furthermore, compounded by their commitments on the other side of the religious divide. The exiled dynasty quite naturally wanted to better the lot of their persecuted co-religionists in the British Isles.[7] Despite constant discrimination and episodically harsh persecution, the English and Scottish Catholics stubbornly cleaved to the Stuarts at least until 1767, when the papacy refused to recognize Charles Edward as rightful king of England, Ireland and Scotland in succession to his deceased father (McLynn 1988, 473–9). Likewise the Irish Catholic community did not entirely abandon hope of a Stuart restoration until the 1790s, and up until the 1760s tens of thousands of young Irishmen took service in the armies of France and Spain at least partially inspired by hopes of restoring the Stuarts and redeeming Ireland from heretic bondage (Ó Ciardha 2002). Hardly surprisingly, the Stuarts felt a corresponding moral obligation towards all three communities. Still more of a problem was the vexed question of how a good Catholic king should deal with heretic subjects. An anonymous memorandum, the 'Grounds of a true Catholick,' drawn up in the early eighteenth century and preserved in the Stuart papers well summarizes the problem:

1. All Catholick good consists in unity.
2. God is simply one.
3. Divine grace, the guifts of the Holy Ghost, all theologicall, cardinal, and morall vertues are simply one in God.

---

[4] [Whitelocke Bulstrode], *A Letter Touching the Late Rebellion, and What Means Led to It; and of the Pretender's Title* (1717), 30.

[5] [Anon], *The Interest of England in Relation to Protestant Dissenters: In a Letter to the Right Reverend, the Bishop of ______* (1714), 27.

[6] John England, *Pray for the Peace of Jerusalem. A Sermon Preached at Sherborne in the County of Dorset on the Publick Fast, March 15, 1709/10, a Little After the Rebellious Tumult Occasioned by Dr Sacheverell's Tryal* (1710), 7.

[7] *Original Papers; Containing the Secret History of Great Britain, from the Restoration to the Accession of the House of Hanover*, ed. James Macpherson, 2 vols (1775), i. 607.

4. All diabolical art tends to violat unity.
5. It is violated by any the least mixture of evill.
6. As the mixture makes it no longer one, so it makes it no longer good.
7. Whersoever the divell has a hand or a finger, evill is certenly thrust in.
8. Of all evills, heresy is the greatest, and most dangerous.
9. No man can love God with supernatural love, unles he hate heresy.
10. No good Catholick can concurre with an heretick, to promote heresy.
11. He that promotes formal hereticks, cannot be said to hate heresy.
12. He that hates not heresy, cannot be a fitt instrument to sett up Catholick religion.[8]

Contemporary Catholic theology thus threatened the Stuarts with moral obloquy unless they personally embraced a theological position that would lead them to initiate policies certain to alienate most of their followers and vindicate claims by Whig zealots throughout the British Isles that restoring the Stuarts would be to impose popery (and doubtless, slavery) on the entire population.

What follows is an analysis of how the exiled Stuarts, and in particular James 'III and VIII' struggled to deal with this conundrum by letters and declarations aimed at his followers, and his putative subjects beyond them, in the early years of his shadow reign. Before delving into this, however, it is worth asking a very central question: in the event of a restoration, how did James in his heart of hearts think he should represent Catholicism in his role as the Catholic king of a Protestant people? The sources offering an answer to this question are more limited than might appear at first sight. The declarations and other formal letters he issued on the subject are in a sense official documents, i.e. statements of official policy. This writer believes he can detect a distinct note of sincerity in these official papers, which after 1709 at the latest were certainly all draughted for, read over, discussed, amended and reamended by James (the self-described 'master of the press') before they were published in France and Italy and smuggled into the three kingdoms.[9] But this is a subjective judgement. What is required to resolve the matter are unguarded words uttered by James, and, leaving aside the distinct tone of anger one might be tempted to detect in some of his official statements, there is a difficulty here in that because of the possible political consequences contemporary princes were brought up not to give vent to their feelings publicly (including in print), and he was no exception (Corp et al. 2004, 257–79).

One interesting statement from someone who was very close to James, may, however, give us a clue as to his personal position on the subject. In July 1713 James's mother, Mary of Modena, who was the former, and highly competent, Queen-Regent of the Jacobite government in exile and remained a close advisor to

[8] Bodleian Library, MS Carte 208, fol. 224a.

[9] HMC *Stuart*, i. 434, 443, 445: James to Bolingbroke, [Bar-le-Duc? and] Commercy, 30 Sept./10 Oct. and 10/21 Oct. 1715; i. 435: Bolingbroke to James, Paris, 7/18 Oct. 1715; i. 449: Bolingbroke's draught of a declaration, amended by David Nairne [Oct. 1715]; i. 455: James to Lewis Innes, 'Larmoisi', 29 Oct./9 Nov. 1715; Lord Mahon, *History of England from the Peace of Utrecht to the Peace of Aix-La-Chapelle*, 3 vols (2nd edn, 1839) i., appendix, xxxvii–xxxviii.

her son after he attained his majority, was on one of her periodic retreats to the convent of Chaillot. There was a good deal of excitement and speculation in Jacobite circles at the time about the possibility of a peaceful restoration of the exiled Stuarts after the death of Queen Anne, who was both ailing and popularly (albeit wrongly) believed to favour her brother's succession (Gregg 1972). This led one of the queen's unnamed companions at the convent to express the hope that James's succession would see the triumph of Catholicism in the British Isles. On the contrary, replied Queen Mary, 'Quand mon fils,...retourneroit, on ne verroit pas pour cela de changement sur la religion; tout ce que l'on pourroit, seroit d'éviter aux catholiques la persécution. La prudence ne permettra pas de rien innover.'[10]

In addition, James in 1716 personally wrote a spectacularly angry response to an attempt to manoeuvre him into publicly suggesting he might have converted (or be about to convert) to Protestantism, and in 1718 he furiously denounced what he perceived as an attempt to lock him into a commitment to return the British Isles to Rome in the event of his restoration. On the first occasion, which arose during James's brief sojourn in Scotland during the 1715 Jacobite rebellion, his advisors had tried to pressure him to attend a Protestant service of thanksgiving for his safe arrival in the country. James refused, pointed to his public commitment to religious toleration, and in a private letter to his leading supporters denounced this attempt to suggest his personal commitment to Catholicism was not rock solid:

> ...since my coming here everybody knows I had not so much as a priest with me nor have not now any living constantly at this place. I hear not Mass so much as every day, and when I do, it is in so private a manner that the last Catholic subject I have could not do it with more caution; and what are the returns I receive for all this, when even that liberty, which in a king would be looked upon as tyranny to refuse to his subjects, is grudged by them to me, who give me in my own person but a sad example of that leniency and moderation in religious matters they preach so much and practice so ill, but which they will never make me desist from showing to them.[11]

The second occasion arose from an alleged attempt by Fr Lewis Innes, Mary of Modena's Almoner, to suggest that James was going soft on Catholic doctrines to please his Protestant adherents. James was outraged, demanded Innes's dismissal, and in a letter to his mother's confessor that he well knew would be circulated at her court at St Germain-en-Laye declared:

> Je suis Catholique, mais je suis Roy, et des sujets de quelque religion qu'ils soient doivent etre egalement protegés. Je suis Roy mais, comme m'a dit le Pape luy meme, je ne suis pas Apotre, je ne suis pas obligé de convertir mes sujets que par l'exemple, ni de montrer une

[10] *Stuart Papers Relating Chiefly to Queen Mary of Modena and the Exiled Court of King James II*, ed. Falconer Madan, 2 vols (Roxburghe Club, 1889), ii. 425: [Early July, 1713]. Queen Mary implied she was willing to make this degree of commitment to the status quo on James's behalf as early as 1702: National Records of Scotland (hereafter NRS), GD 26/8/139: Secretary of State Sir Charles Hedges to Chancellor James Ogilvy, earl of Seafield (copy), Whitehall, 19 May 1702 (this is an informer's account of negotiations with the Jacobite Court by John Hamilton of Biel, baron Belhaven, on behalf of the Duke of Hamilton).

[11] HMC, *Stuart* iv. 13: James to [his Council?], [Perth] Jan. 1716.

partialité apparente aux Catholiques, qui ne serviroit qu'a leur nuire effectivement dans la suitte. Tout le monde peut scavoir et mes sentiments et ma conduite egalement juste et honorable sur ces chefs et ce n'est qu'une malice interessée qui puisse le faire paroitre dans un faux jour, mais avec un tel secours je ne m'etonneray jamais en voyant que les Protestans seuls me rendent justice en certaines choses, et que les Catholiques seduits par ceux, qui feignant la probité et le zele, pouvant accuser d'hypocrisie et de lacheté une politique egalement necessaire, prudente et Catholique.[12]

The upshot of all of which is that it seems likely that James genuinely accepted that in the event of a Stuart restoration he would not, and could not, actively seek to return the three kingdoms to Rome. He could, and implies he would, set a personal example that might inspire voluntary conversions, but he (and the dynasty) would be ruling through Protestant ministers and a Protestant bureaucracy for the foreseeable future and the best the Catholics of the British Isles could hope for was to be allowed secluded, private worship (as would he) and an end to officially-instigated bouts of persecution.[13]

While not ideal from a Protestant Jacobite point of view, such a religio-political settlement would probably have been acceptable. James's problem then, was twofold. He had to persuade the Protestant Jacobites that he was sincere in his commitments, and thus that Protestantism in the British Isles would be secure if he was restored, and that this was the best deal they were going to get from him. This latter stemmed from the widespread hope among the Protestant Jacobites that if only James heard the Protestant message preached he would convert, and their general belief that Charles II had been a secret Catholic long before his deathbed conversion in 1685 (Szechi 1984, 37). The corollary of Charles II's having allegedly dissembled most of his life was that James, like him, could either obtain secret permission from the pope to dissemble and pretend he was a Protestant, or just fake it anyway. In addition, and compounding the difficulty of achieving the above, James had to stay in conformity with Catholic teaching on the subject of not tolerating evil, of which, as we have seen, heresy was rated as one of the worst examples.

The process of getting the core message over to James's Protestant followers began soon after he succeeded his father in 1701. Since James was only 14 years old at the time, and subject to a regency council headed by his mother and Charles Middleton, earl of Middleton, the, at that time Protestant, premier Jacobite Secretary of State, we must presume that this first statement of his intentions towards the Protestant Churches

---

[12] HMC *Stuart*, v. 515: James to Fr Honoré Gaillard, Fano, 17/28 Feb. 1718.

[13] In contrast to the interpretation advanced here it must be acknowledged that in 1715 James signed a secret agreement with the Spanish government promising to do his best to restore the British Isles to Catholicism (L. B. Smith 1982, 163). This, however, had more to do with Spanish politics than Jacobite ambitions. To receive promised financial support from Spain in 1715 the future King James was required to promise to deliver something meaningful in a Spanish political context. There is also a world of difference between doing one's best within the restrictions of the law and the realms of the possible and being willing to do anything necessary to achieve a particular result. Hence, in the opinion of this writer, James's conduct and statements in other contexts are more significant than his secret treaty commitments to Spain.

of the British Isles was primarily generated by others. It was certainly in conformity with the policy imposed by the Protestant party within the Jacobite movement, who had defeated their Catholic party rivals and established an alternative Jacobite Revolution Settlement at the heart of Jacobite ideology in 1693 (Szechi 1993). This had initially only been reluctantly accepted by James II and VII, though he later came to embrace it more positively.[14] The first set of 'Instructions' issued in James III and VIII's name, though, sent as a manuscript open letter to those in England in regular receipt of correspondence from St Germain (who then copied and distributed it throughout the English Jacobite community), went notably further in terms of what it committed him to doing should he regain his throne, and they are worth citing in full to establish the general tenor and basic extent of James's public commitments with respect to the Church of England and its sisters for the rest of his reign:

> You shall acquaint our friends, that when it shall please God to put us in possession of our kingdom of England, we promise that we will govern according to law, and that we will secure and protect all our subjects of the Church of England as it is established by law, in the full enjoyment of all their legal rights, privileges, and immunities, and in the sole possession of all their churches, universitys, colleges, and schools; and that in all vacancys of bishopricks and other dignities or benefices at our disposal, care shall be taken to have them filled with the worthiest members of their own communion.
>
> And, if it shall appear, that by their loyal endeavours and assistance, those of the church of England have been instrumental in our restoration, we further promise, that in their behalf and for their greater security, we will so far wave during our own reign our right to nomination to bishopricks and all other dignitys and benefices in the disposal of the crown, that we will appoint the archbishop of Canterbury for the time being, and four bishops, to propose three persons for each vacancy, of which number we shall chuse one; but this without prejudice, for the future, to the undoubted right of the crown.
>
> We are also, upon the same condition, willing to remit, during our reign only, all the tenths and first fruits paid by the bishops and clergy to the crown, to those who shall return to their duty and repair their fault by endeavouring to reclaim their misled flocks. But this our bounty is not to be extended to those who perversely persist in their error; and who, by renouncing the principles of the church of England have no pretence to the benefit we intend her. And this concession shall not be made a precedent nor drawn into consequences against our lawful successors.
>
> As on one side, we solemnly promise to govern by law and inviolably to maintain the liberty and property of our subjects, so on the other side, we would not be understood to lye under any obligation of persecuting those of our own religion or any other dissenters merely upon the account of conscience. And we shall leave it to our first Parliament to agree upon and settle a just and equitable moderation of the laws now in force against Roman Catholicks.[15]

This was *prima facie* a strong and sweeping commitment to preserving the Protestant hegemony in the British Isles, while at the same time making it clear that were he to be in power at Westminster James would seek to extend religious toleration to the Roman Catholic community and shield them from further persecution.

[14] Westminster Diocesan Archive, Old Brotherhood MSS, III pt 3, ep. 259: memo of a conversation with James II and VII by Sir Edward Hales, [St Germain] 23 May/2 June 1693.

[15] *Original Papers*, i. 606–7: 'Copy of his Majesty's Instructions sent into England', 20 Feb./3 Mar. 1702.

The same basic commitments were to be made in declaration after declaration, letter after letter, for the rest of James's life. Though the very first draught and its accompanying editing process may not have been overseen by him, the lack of fundamental change in the principles expressed above suggests that he was in agreement with the central thrust of the 'Instructions', and the only significant changes that followed as he grew into his role as king-in-exile were responses to particular religious crises in the British Isles.[16] Thus at the height of the Bangorian Controversy in 1717, for example, James made it clear in a 'personal' letter to Dr Charles Leslie (the most well-known and respected Protestant clergyman in exile) that he knew would be transmitted to his Protestant followers in the British Isles, that, should he be restored, he would uphold the authority of the Church of England to discipline its clergy and that of Convocation more generally:

> ...by the best information I can have the intrinsic spiritual power of the Church, or power of the Keys as exercised by the Apostles and most pure and primitive Church in the first three centuries has ever been thought an essential right of the Church of England, so that it may inquire into the doctrines of its own members and inflict ecclesiastical censures, not extending to any civil punishment. Now the civil government's putting a stop to such proceedings is in effect taking away that undoubted right of the Church, which, if it please God to restore me to my own just right, I am firmly resolved to maintain to it.[17]

But otherwise he simply referred them to his previous declarations and commitments.

The 'Instructions' were, however, very carefully constructed. In 1716 Father Lewis Innes and Dr John Ingleton were asked for their formal, theological opinion as to whether James could legitimately, as a good Catholic, promise, 'to protect and maintain the Church of England as established by law', and their response clearly sets out the theological reasoning underpinning all of James's declarations going back to the 'Instructions'. The two forthrightly rejected any such wording:

> We are of opinion that he cannot with a safe conscience make any promise in those words. We ground our opinion on these principles, which seem to bear no dispute.
>
> 1. 'Tis lawful to permit and tolerate evil in some cases, but never to approve it.
> 2. All expressions which argue any more than a permission reach to an approbation; and all approbation or consent to evil is unlawful.
>
> Now these words protect, or defend and maintain certainly express or imply more than a permission, and therefore are not warrantable, for how can any man promise to protect and maintain which in his conscience he condemns and abhors? Hence no man can promise to protect and maintain robbery, usury, etc., because this would be to approve or consent to evil, and more especially heresy established by law is a far greater evil than these.[18]

---

[16] HMC *Stuart*, i. 218–20: 'James VIII to his good people of his Ancient Kingdom of Scotland', St Germain, 18/29 Feb. 1708; i. 343–5: Declaration for England, [Bar-le-Duc?] 3/14 Jan. 1715; Royal Archives, Windsor, Stuart Papers, Box 6/26: 'His Majesty's Most Gracious Declaration,' Pesaro, 10/21 Mar. 1717.

[17] HMC *Stuart*, v. 244: Urbino, 18/29 Nov. 1717.

[18] HMC *Stuart*, ii. 187: formal opinion by Lewis Innes and Dr John Ingleton on whether James III and VIII can promise, 'to protect and maintain the Church of England as established by law', 15/26 May 1716.

They then turned their attention to what James could faithfully promise without violating the precepts of his own religion. Taking it as a given that James had received indications from the Protestant Jacobites that what they wanted was assurances that he would not, as king, molest them or alter anything associated with the religious settlement, Innes and Ingleton posited that if he promised, 'that I will not alter the religion established by law, nor change the laws relating thereunto, nor will I molest the professors of it, but on the contrary protect and maintain them in all their just rights, dignities, privileges and possessions', he would not breach Catholic doctrine on the subject.[19] Their reasoning was founded firmly in rational pragmatism: 'It will never be in the King's power to change these laws without a Parliament; consequently he may promise not to do it, for any man may promise not to do what he really cannot do.'[20]

The wording of the statements that came out of this periodic draughting process was accordingly very careful. It implicitly skirted the question of whether or not any of the Protestant Churches of the British Isles were true Churches or simply collections of individuals, avoided the question of whether or not Protestants could be pious, good individuals and where they were likely to go after they died, and maintained a discreet silence on the question of whether James would promote the Protestant Churches' interests.[21] Such apparently weaselly semantic equivocation was regarded with deep suspicion by contemporaries, and it is certainly possible to read his statements in this way, but taken in context it seems more likely that James and his ministers were simply trying to square the circle.[22] For a host of personal and diplomatic reasons James was not going to stray beyond the precepts of Catholic orthodoxy in his public statements. But this should not be taken to mean that he intended to betray the spirit of the commitments he entered into after 1702. And in this context it is worth remembering that at some risk to his reputation for being a good Catholic James quietly sheltered Protestant ministers and facilitated Protestant religious services while living as a pensioner of the pope himself in Avignon, Pesaro, Urbino and Rome (Corp 2009, 19, 51).

It is also worth noting that there was, too, another weighty opinion on the subject of how far Catholic princes could legitimately sponsor or allow Protestant religious practice than that advanced by Innes and Ingleton. When James II and VII in late

[19] HMC *Stuart*, ii. 187: formal opinion by Innes and Ingleton, 15/26 May 1716.

[20] HMC *Stuart*, ii. 188: formal opinion by Innes and Ingleton, 15/26 May 1716.

[21] HMC *Stuart*, i. 449: Bolingbroke's draught of a declaration, with amendments by James's confidential secretary for Catholic correspondence, David Nairne [Oct. 1715].

[22] Mahon, *History*, i., appendix, xxxvii: Bolingbroke to James, Paris, 22 Oct./2 Nov. 1715; Sir Richard Steele, *A Defence of the Crisis, Written by Mr. Steele. Containing, a Farther Vindication of the Late Happy Revolution. And the Protestant Succession to the Crown of England, in the Illustrious House of Hanover* (1714), 22–23, 25. Eighteenth Century Collections Online. Gale. University of Manchester – John Rylands. 9 Sept. 2010.

http://find.galegroup.com/ecco/infomark.do?&contentSet=ECCOArticles&type=multipage&tabID=T001&prodId=ECCO&docId=CW3304282530&source=gale&userGroupName=jrycal5&version=1.0&docLevel=FASCIMILE.

1692 finally, reluctantly accepted that he was going to grasp the nettle and commit himself to the Protestant party's agenda he salved his conscience by submitting the text of the pivotal royal declaration to two panels, one of English and Scots theologians (including Lewis Innes) and one of French theologians. The Anglo-Scots panel roundly rejected any wording that promised to 'protect and defend', but allowed that he could promise to protect and maintain the Protestant Churches. The French panel took a much more radical line and authorized James II and VII to make far more sweeping commitments, which he did (and initially regretted, though he did not try to go back on them).[23] The same split opinion was still to be found in 1716, when Innes and Ingleton's view on the subject ran directly contrary to that of Doctor of the Sorbonne Louis Ellies Dupin (a leading Gallican theologian), who condemned any attempt to alter the established religion in the British Isles on the grounds that it flew in the face of God's providence.[24] James cleaved rather to the conservative than the radical side in response to these bodies of opinion, but sought to underscore his commitment to maintaining a Protestant settlement as emphatically as possible within the bounds set by Innes and Ingleton, as may be seen in his next public declaration on the subject after these exchanges:

> Wee do likewise renew and confirm, all the Promises made by Us in Our foresaid Declarations, to protect, support and maintain Our Subjects of the Church of England and Ireland in the full and free exercise of their Religion, and to secure the said Church as by Law established, and all the Members thereof, in as full enjoyment of all their Legal Rights, Privileges and immunities, and in as full and peaceable possession of all their Churches, Universities, Colleges and Schools, as ever they enjoyed them under any of our Royal Predecessors of the Protestant Communion.[25]

In contemporary terms this was about as far as a genuinely Catholic king could go in terms of an express, public commitment to maintain the Protestant religious settlement in the British Isles, but it was still not enough to remove the suspicions of the Protestant Jacobites.

The roots of their seemingly inerasable apprehensions went back to the reign of James II and VII. He had come to the throne in 1685 on a tide of Tory enthusiasm. As far as they were concerned at that time they had a workable deal with the king, which amounted to him being allowed freely to practice his religion and install Catholics in his household, and protect the Catholic community from persecution by various legal sleights-of-hand, on condition that he basically left the military, the Church and the legal scaffolding guaranteeing Protestant hegemony throughout the British Isles alone (Harris 2007, 40–61, 196–7). In the expectation that he would be succeeded by one of the Protestant children of his first marriage (this was the lowering cloud on the king's horizon until the birth of the future James 'III and VIII'

---

[23] *The Life of James the Second, King of England, &c. Collected out of Memoirs Writ of his own Hand,* ed. J. S. Clarke, 2 vols, (1816), ii. 508–10.

[24] Scottish Catholic Archives (thereafter SCA), BL 2/210/6: [David Nairne] to Thomas Innes, [Chaillot] 22 July/2 Aug. 1716.

[25] Stuart Papers, Box 6/26: 'His Majesty's Most Gracious Declaration,' Pesaro, 10/21 Mar. 1717.

on 10 June 1688), however, James II and VII succumbed to the temptation to try and create 'facts on the ground' before he died. From the autumn of 1685 he therefore pursued a broad policy of Catholicisation at breakneck speed. This involved the deliberate packing of the bench to secure favourable legal decisions allowing him to dispense with and suspend laws he did not like, a systematic attempt to stuff Parliament with compliant M.P.s, purges of the military and officialdom to install Catholics willy-nilly and infringement of the liberties and privileges of the university colleges to try and establish a Catholic presence at the heart of contemporary intellectual life (Harris 2007, 2–3, 187–235, 239). These measures contributed directly to the alienation of the majority of the Protestant population in England and Scotland by the summer of 1688, and hence to James II and VII's deposition and William III and II's successful usurpation during the winter of 1688–1689 (Speck 1988, 71–116). Moreover, and very significantly for the internal politics of the Jacobite movement, James II and VII's conduct when in power was bitterly remembered by virtually all elements of the Protestant community in the British Isles—including the Protestant Jacobites whose agenda dominated the movement after 1693.[26] In consequence, despite the sweeping constitutional concessions he (probably sincerely) promised, James II and VII was dogged by a deep well of suspicion, even on the part of the Protestant Jacobites, for the rest of his life.[27]

When he succeeded the old king in 1701 James, Prince of Wales in many ways represented a new start for Jacobitism, but it took him nearly 20 years to shake off his father's legacy. In consequence the Protestant Jacobites episodically badgered him over two issues: cast-iron constitutional concessions that would guarantee that if he was restored he would not be able to repeat his father's assault on the Protestant hegemony and his own conversion to Protestantism.

There were a range of measures proposed with respect to ensuring James's future constitutional good behaviour. The least explicit stemmed from the English Jacobites, who appear basically to have been relying on a raft of unspecified early legislation in a specifically free Parliament which they confidently expected to dominate.[28] It is legitimate to assume that they intended securely to shackle a future King

[26] Westminster Diocesan Archive, AAW, B6, item 21a: account of Richard Grahme, viscount Preston's trial at the Old Bailey [in Roger Palmer, earl of Castlemaine's hand], 17 Jan. 1690; AAW, B6, item 244: anon memo, 16 Dec. 1690; Old Brotherhood MSS, III pt. 3, ep. 231: copy 'Sir William Ellis's paper on occasion of the late earthquake', [1692?]; La Courneuve, Archives étrangères, Correspondance politique (Angleterre; thereafter AECP (A)) 251: Ambassador Charles François de la Bonde, marquis d'Iberville, to Jean Baptiste Colbert de Croissy, marquis de Torcy, [London] 10 Apr. 1714.

[27] See for example Middleton's reaction to James II and VII telling him that he only ever intended to ensure there should be one Catholic college at Oxford: Old Brotherhood MSS, III pt 3, ep. 259: [Addendum in Hales's hand but unsigned] 2/13 Oct. 1694 (n.b. in the collection's catalogue the 'M' referred to in the addendum is misidentified as John Drummond, earl of Melfort; in fact 'M' is certainly Middleton).

[28] MS Carte 180, fol. 3r: 'His Majestie's Most Gracious Declaration to all his Loving Subjects of his Kingdom of England', Bar-le-Duc, 9/20 July 1715; HMC *Stuart*, i. 448: Declaration for England by James (draughted by Bolingbroke), Commercy, 14/25 Oct. 1715.

James to a solidly Protestant settlement, but since they never revealed any specifics it is impossible to say exactly how they intended to do so. At the other end of the range were the Scots Jacobites. Because Scotland had a far stronger, explicit and legally binding constitutional tradition than England they were much more forthright on the subject, and nowhere more so than in their secret negotiations with France and the Jacobite Court between 1705 and 1708.[29]

These aimed at agreeing the terms and conditions under which an army of Jacobite Scots would rise for James and against the Union with England in the event of a French invasion. As a precondition the Scots Jacobite leadership demanded by letter and by formal memoranda covertly sent into France that James accept a new constitution embodied in the 'Heads of the Instrument of Government'. This stipulated,

> That no Papist be imployed in places civil, ecclesiasticall, or militarie, without the consent of Parliament.
>
> That no alyance be made, either public or private, without consent of Parliament.
>
> That the officers of state, privie councell, lords of session and justiciarie, be named by Parliament.
>
> That a Parliament be caled and continowed to sitt for the dispatch of publick business and the service of the nation once in three years.
>
> That a libertie of consitience and toleration be setled upon a right foundation.[30]

And in an associated document (the 'Scheme') they further required,

> That the scheme of the securitie be form'd in limiting the prerogative of a Popish King and establishing a clame of right and instrument of government containing the essentiall and warantable conditions by which a prince of that comunion is to be regulat under a failor of the aledgiance of the subject.
>
> That a convention be caled, and the instrument of government concerted and adjusted before the King enter upon the administration of the government.
>
> That the instrument be confirm'd by the solemn oath of the Prince and the states, bearing a declaration against all .[blank]. or mental reservation as uterlie inconsistant with human societie, morality, or religion.
>
> That the oath of aledgiance be concerted with relation to the instrument of government.[31]

It is not going too far to say that such radical constitutional demands (to which, as far as we can tell—there are gaps in the sources—James apparently agreed) would, if implemented, certainly have shackled him, and in fetters of steel. The Scots at this time, though, were in a hurry and dead set on securing prompt French

---

[29] *Correspondence of Colonel N. Hooke, Agent From the Court of France to the Scottish Jacobites, in the Years 1703–7*, ed. W. D. Macray, 2 vols (1870), ii. 330–31: memo on a restoration by James Grahme of Newton, June 1707; ii. 332–3: memo of reasons for the restoration of James VIII by Anne Drummond, dowager countess of Erroll, June 1707; ii. 335–6: [statement by the Jacobite 'Juncto'] June 1707. That such an approach to the problem of the constitutional relationship between a Catholic king and a Protestant people was being mooted from the beginning of James's reign as Jacobite king may be seen from: NRS, GD 26/8/139: Hedges to Seafield (copy), Whitehall, 19 May 1702.

[30] *Correspondence of Colonel N. Hooke*, ii. 335.

[31] *Correspondence of Colonel N. Hooke*, ii. 333–4.

military support, and so were willing to overlook the other option for ensuring Protestant security under a future King James: his conversion to Protestantism.

From the Protestant Jacobites' point of view this would have been the ideal remedy, as may be seen from persistent rumours and stories bruited about amongst them that James had in fact converted to Protestantism.[32] At a stroke it would have removed the basis for all their apprehensions about the future conduct of the exiled dynasty. This inclination (actually closer to fixation) lay relatively dormant while the only immediate prospect of James's restoration was by force of arms, and in particular, Catholic French arms.[33] But by 1710 France's worsening military crisis made it clear that this was not going to happen any time within the foreseeable future, and that the best hope of a restoration now lay in James's negotiated succession to Queen Anne. The Protestant Jacobites correspondingly became alert to every opportunity they could find to broach the issue and, if at all possible, pitch the message: convert and your restoration will be guaranteed. In general, this was just hinted at, as in 1711 when in a general letter directed to Middleton 'Lord Lumb' (probably Dr Leslie) opined: 'Je ne vous rien cacher, la seule objection contre le Roy est celle de sa religion, et cela ne luy est pas inpute comme sa faute, mais comme son malheur et le notre'.[34] Or in 1712, when an anonymous group of Jacobite Parliamentarians wrote to St Germain to ask that he hear the Protestant message preached immediately, but promised that if he was not persuaded they would drop the matter.[35] Yet at the same time they forthrightly tried to close off any option but conversion.[36] By 1713 the Protestant Jacobites were surreptitiously writing to Protestant courtiers at St Germain to encourage them to raise the issue, and when James ostentatiously invited Leslie over to minister to his Protestant servants at his new residence in Bar-le-Duc Leslie took the first opportunity to raise the matter with him.[37] This campaign came to a crescendo early in 1714, when Queen Anne's ministers Robert Harley, earl of Oxford, and Henry St John, viscount Bolingbroke, for their own (tactical political) reasons sent messages to James via a French envoy, the Abbé François Gaultier, informing him: 'that without he reforms actually or gives hopes that he will, his best friends are resolved to abandon his cause, it being impossible

[32] SCA, BL 2/173/10, 16: Fr James Carnegy to the Scots College, [Edinburgh?] 3 June and 16 Aug. 1712; SP 54/9/3b: copy of 'A Letter from a Gentleman in the Earl of Mar's Camp to his Friend in the West Country,' 1 Oct. 1715; AECP 265, fol. 270v: d'Iberville to Torcy, London, 27 Jan./7 Feb. 1716.

[33] That the idea was nevertheless lurking in the background of Protestant Jacobite thinking is apparent from the informer's account of Belhaven's conversations with Mary of Modena: NRS, GD 26/8/139: Hedges to Seafield (copy), Whitehall, 19 May 1702.

[34] MS Carte 180, fol. 293: Memo on projects and affairs in England by Lord Lumb, April, 1711.

[35] AECP (A) 242, fol. 137: [Memo from St Germain to Torcy, c. Oct 1712?].

[36] AEMD (Archives étrangères, Mémoires et documents) 138: 'Memoire sur le Genie des Anglois, et sur l'Estat present de la cour d'Angleterre', 29 Oct. 1712.

[37] HMC *Stuart*, i. 252, 259–60, 273: Berwick to James, St Germain and Fitzjames, 9/20 Nov. 1712 and 8/19 Mar. and 11/22 Aug. 1713.

for them to help him otherwise, and that they will not ruin themselves *gratis*.'[38] These two were then reinforced by a cascade of further verbal messages (including some from prominent Catholics), transmitted in official correspondence by intermediaries such as the French ambassador to Britain, and personal letters from genuine Jacobites, all demanding James convert on pain of abandonment by his friends in England.[39]

If James would not convert, and very much as a second best, the Protestant Jacobites hoped that he could be persuaded to dissemble conversion while secretly remaining a Catholic. As one old Protestant courtier told Queen Mary at Chaillot in 1713, 'la nécessité des tems exigeoit des condescendances; que le prophête Elizée qui ne manquoit ni de zèle ni de lumière n'avoit point empêché que Naaman après sa guérison ne suivit son Roi dans le temple Remnon.'[40] Oxford and others slyly pressed him to do this later the same year, when (again via Gaultier) they urged James to 'imite la conduite de Charles Second son oncle'.[41] When James and his advisers stubbornly refused to take the hint,[42] more direct pressure was applied: 'il est absolument necessaire', Gaultier wrote to James in February 1714, 'que vous dissimuliez vôtre Religion ou que vous la changiez entierment.'[43] Even Leslie, who disliked the idea of dissimulation, privately pleaded with James to allow him to spread the word that James would 'examin and consider' the Protestant case, and thereby imply that James was on the right path.[44]

James, however, would have none of it. Though he was more or less willing to do other things to placate his importunate followers, such as banishing the Jesuits they so hated from his court, establishing (as soon as he could) a Protestant chapel with a regular minister for his servants and reluctantly replacing his closest adviser, the now-Catholic Middleton, with the Protestant (but sadly incompetent) Thomas Higgons, he would go no further.[45] Back in 1711, in a letter in his own hand (again

---

[38] University of Pennsylvania, MS French 139, fols 265–85: Abbé François Gaultier to Torcy, [London] 26 Feb. 1714; AECP (A) 251, fol. 126: d'Iberville to Torcy, [London] 18 Feb. 1714; SCA, BL 2/191/8: Lewis Innes to Thomas Innes, [Bar-le-Duc] 23 Jan./3 Feb. 1714.

[39] SCA, BL 2/191/6, 8, 9: Lewis Innes to Thomas Innes, [Bar-le-Duc] 16/27 Jan., 23 Jan./3 Feb. and 28 Jan./8 Feb. 1714; AECP (A) 251, fols 148, 209: d'Iberville to Torcy, [London] 26 Feb. and 23 Mar. 1714; Westminster Diocesan Archive, Ep. Var. V, ep. 48: Edward Dicconson to [Laurence Mayes], [Douai] 16/27 Mar. 1714.

[40] *Stuart Papers Relating Chiefly to Queen Mary of Modena*, ii. 438 [26 July/6 Aug. 1713]. The Biblical reference here is to the tacit permission given to the convert Syrian general Naaman by the prophet Elisha to accompany his (pagan) king into the temple of Rim and there bow before the idol alongside him; II *Kings*, 5.

[41] L. G. Wickham-Legg, 'Extracts from Jacobite Correspondence 1711–14,' *English Historical Review*, 30 (1915) 506: Gaultier to Torcy, London, 3/14 Dec. 1713.

[42] University of Pennsylvania Microfilm, MS French 120/6: Torcy to Gaultier, 8/19 Dec. 1713.

[43] Wickham-Legg, 'Extracts,' p. 508: Gaultier to [James], London, 6 Feb. 1716.

[44] SCA, BL 2/191/10: Lewis Innes to Thomas Innes, [Bar-le-Duc] 6/17 Feb. 1714.

[45] SCA, BL 2/173/10: Carnegy to the Scots College, [Edinburgh?] 3 June 1712; HMC *Stuart*, i. 260: Berwick to James, St Germain, 17/28 Mar. 1713; Szechi 1984, 7, 187–8.

designed for wider manuscript copy and distribution in the British Isles) he had signalled his willingness to hear the Protestant case once he was restored,[46] but in the interim he absolutely refused to dissimulate under any circumstances:

> I neither want counsell nor advice to remain unalterable in my fixed resolution of never dissembling my religion, but rather to abandon all th[a]n act against my conscience and honour, cost what it will. These are my sentiments, and had I others, or should I act contrary to those I have where is the man of honour that would trust me? And how could my subjects depend upon me or be happy under me, if I should make use of so notorious an hipocrisy to get my self amongst them?[47]

This response, backed up by a pamphlet written by Leslie explaining James's reasoning at length was, as far as James was concerned, his last word on the subject.[48] And though it did not completely shut down the issue, nor finally eliminate Protestant Jacobite fantasies that he had, or might be persuaded to, convert, it did begin the process where it once again became dormant, at least until the birth of James's son, Charles Edward.[49] By 1718 James could privately reassure Cardinal Filippo Antonio Gualterio that, 'Il n'est plus question, Dieu mercy, de ma religion personelle: on a perdû toute esperance de changement.'[50]

So how successful was James's presentation of the prospect of tolerant Catholic kingship to his Protestant audience? *Prima facie* it looks like a failure. By pursuing him so stubbornly on the subject of conversion or dissimulation the Protestant Jacobites were strongly signalling their enduring suspicion of Catholicism and their scepticism about the king's own trustworthiness in his (hoped-for) future role as Catholic king of a Protestant people. Looking a little deeper, however, changes the picture. James succeeded in both staying on the right side of Catholic orthodoxy in his public and private statements and in keeping the Protestant Jacobites on board despite their disappointment. There were, of course, many reasons why the Protestant Jacobites continued to cleave to him, but we should not ignore his (mostly) patient, persevering reassurance of them on the subject of a future religious settlement. Though James was without doubt a rock-solid Catholic, and thus bound to be suspected of ulterior motives by a great many of his followers, he regarded himself as, 'born to live in a Protestant country...among Protestant subjects'.[51] In his formal and informal public statements he correspondingly projected a moderate, tolerant approach to Catholic kingship that in the end calmed his Protestant adherents down and persuaded them (by and large) to accept his personal religious convictions.

[46] MS Carte 210, fol. 409r: [holograph abstract, St Germain] 21 Apr./2 May 1711.

[47] MS Carte 210, fol. 409v: abstract of James III's holograph reply to the English Jacobites, [Bar-le-Duc] 2/13 Mar. 1714.

[48] Stuart Papers Box 6/6: *A Letter From Mr Lesly to a Member of Parliament in London* (1714).

[49] AECP (A) 270, fol. 94v: d'Iberville to Torcy, London, 26 Sept. 1715 and AECP (A) 265, fols 251v: d'Iberville to Torcy, London, 10/21 Jan. 1716; HMC *Stuart*, vi. 166: Charles Boyle, earl of Orrery, to John Erskine, earl of Mar, 6/17 Mar. 1718; Cruickshanks (1979), 13.

[50] HMC *Stuart*, vi. 134: James to Gualterio, 28 Feb./11 Mar. 1718.

[51] HMC *Stuart*, vi. 134: [memo for Gualterio in James's own hand, 28 Feb./11 Mar. 1718].

Perhaps the greatest testimony to his achievement lay in the character and vocation of one of the most important converts to the Stuart cause in the late 1710s: Francis Atterbury, bishop of Rochester. By 1718 the zealously Anglican Atterbury was willing to risk being hanged, drawn and quartered to try and restore James III and VIII to the thrones of the three kingdoms. As far as he was concerned, Catholic King James had become a far better guarantor of right religious order in the British Isles than Protestant King George.

References

Bohanan, Donna. 1992. *Old and new nobility in Aix-en-Provence 1600–1695*. London: Louisiana State University Press.
Bonney, Richard. 1991. *The European dynastic states 1494–1660*. Oxford: Oxford University Press.
Corp, Edward. 2009. *The Jacobites at Urbino: An exiled court in transition*. Basingstoke: Palgrave.
Corp, Edward, with Edward Gregg, Howard Erskine-Hill, and Geoffrey Scott. 2004. *A court in exile: The Stuarts in France, 1689–1718*. Cambridge: Cambridge University Press.
Cruickshanks, Eveline. 1979. *Political untouchables: The Tories and the '45*. London: Duckworth.
Gregg, Edward. 1972. Was Queen Anne a Jacobite? *History* 57: 358–75.
Harris, Tim. 2007. *Revolution: The great crisis of the British monarchy, 1685–1720*. London: Penguin.
Hubatsch, Walther. 1975. *Frederick the Great: Absolutism and administration*. London: Thames and Hudson.
McLynn, Frank. 1988. *Charles Edward Stuart: A tragedy in many acts*. London: Routledge.
Miller, John. 1977. *James II: A study in kingship*. Hove: Wayland.
Monod, Paul Kléber. 1989. *Jacobitism and the English people, 1688–1788*. Cambridge: Cambridge University Press.
Ó Ciardha, Éamonn. 2002. *Ireland and the Jacobite Cause, 1685–1766: A fatal attachment*. Dublin: Four Courts Press.
Smith, L.B. 1982. Spain and the Jacobites, 1715–16. In *Ideology and conspiracy: Aspects of Jacobitism, 1689–1759*, ed. Eveline Cruickshank, 159–78. Edinburgh: John Donald Press.
Speck, W.A. 1988. *Reluctant revolutionaries. Englishmen and the revolution of 1688*. Oxford: Oxford University Press.
Szechi, Daniel. 1984. *Jacobitism and Tory politics, 1710–14*. Edinburgh: John Donald Press.
Szechi, Daniel. 1993. The Jacobite revolution settlement, 1689–1696. *English Historical Review* 108: 610–28.
Szechi, Daniel. 1994. *The Jacobites: Britain and Europe 1688–1788*. Manchester: Manchester University Press.
Szechi, Daniel. 2010. Jacobite Scotland and European great power strategy. *Northern Scotland* 1: 42–53.

Chapter 8
'Every Time I Receive a Letter from You It Gives Me New Vigour': The Correspondence of the Scalan Masters, 1762–1783

Clotilde Prunier

In 1560, the Scottish parliament outlawed Catholicism. As a result, the now clandestine Roman Catholic Church had to resort to makeshift arrangements not to be altogether wiped out of the country. One of the most pressing emergencies was to contrive a means of training Scottish-born priests, as it was of the utmost importance that there should be enough clergy to cater to the religious needs of those who had remained staunch in the Catholic faith.[1]

In the decades following the Reformation, Colleges were founded on the Continent, purposefully to maintain Scottish boys training for the priesthood. In the early eighteenth century, the four Scots Colleges were still the Scottish Mission's main source of priests. The Vicar Apostolic, James Gordon, deemed the supply inadequate in many ways, and decided to establish a seminary on Scottish soil. In spite of the tightening grip of the law, which particularly forbade Catholics to teach, and the treatment meted out to priests in the aftermath of the '15, a seminary was established in 1716 at Scalan, a secluded spot near Chapeltown in the Braes of Glenlivet. Scalan was deemed a safe place enough to settle a seminary, as it was quite difficult of access. Further, it was located on the estate of the (Catholic) Duke of Gordon, whose influence ensured that the numerous Catholics living on his lands were out of harm's way. The original building was burned down in the aftermath of the '45 and a smaller, makeshift, one was erected in 1747. The foundation stone of the seminary, as it stands now, was laid in 1767.[2] On its foundation, it was meant to

My grateful thanks are due to the Catholic Record Society which awarded me grants out of the David Rogers Research Fund.

[1] On the Catholic Church in Scotland in the eighteenth century, see Johnson 1983; Szechi 1996; McMillan 1999 and Prunier 2004.

[2] For an account of the seminary, see Johnson 1983 p. 60–70 and Watts 1999.

C. Prunier (✉)
Département d'études anglophones, Université Paris Ouest Nanterre La Défense,
200 avenue de la République, 92001 Nanterre cedex, France
e-mail: cprunier@u-paris10.fr

A. Dunan-Page and C. Prunier (eds.), *Debating the Faith: Religion and Letter Writing in Great Britain, 1550-1800*, International Archives of the History of Ideas 209,
DOI 10.1007/978-94-007-5216-0_8,

serve Scotland as a whole, but became the Lowland seminary when the Mission was divided into the Highland and Lowland Districts in the early 1730s. It was hoped Scalan would bring a welcome addition to the Scottish Catholic clergy, providing the Mission with 'heather priests', that is to say clergymen who had received their whole training in Scotland, rather than abroad, at the hands of masters conversant with the peculiar demands of the local situation. There can be no doubt that this was Bishop Gordon's design, as George James Gordon reminded Bishop Gordon's successor, Alexander Smith, in 1763, while he lamented the new master's apparent resolution to 'turn Scal: into a kind of nursery for the forreign Shops; directly opposite to the wise intention of Mr Nicop: the founder'.[3] It is no coincidence that the man who insisted Scalan retain its original purpose was himself the very first 'heather priest' ordained at Scalan and a former master there. Indeed, so intense were his feelings for his 'old Nest', that even after he left it to serve the mission of Aberdeen in 1739, he went on to subscribe himself, and was referred to, as 'Scalanensis' until he died in 1766.[4] However, he could not but know that he was one of only four 'heather priests' bred at Scalan in almost half a century.[5] Further, some of those boys sent to Scots Colleges had never attended Scalan at all, but had been tutored by their local priests, before pursuing their studies abroad. That happened to be the case of John Geddes, the new master appointed to Scalan in 1762.

The tensions between these two men's views on the function and uses of Scalan are apparent in their correspondence with Bishop Smith. Indeed, except for the annual meeting of the Bishops and the administrators of the Mission,[6] there were very few opportunities for face-to-face encounters, so that letters were by force of circumstance the main vehicle for all such exchanges of opinions among the clergy of the underground Church. In the late eighteenth century, correspondents still made use in their correspondence of aliases and of various metaphors, in particular commercial ones. For example, 'labourers' (ecclesiastics) catered for the needs of their 'customers' (the faithful) and 'apprentices' (prospective priests) were sent to 'shops' (Scots Colleges) to learn their trade.[7] However, by then, most letters were sent

---

[3] Scottish Catholic Archives, Blairs Letters [thereafter SCA, BL], George James Gordon to Alexander Smith, 7 March 1763. The foreign shops were the Scots Colleges on the Continent; 'Mr Nicop' was one of Bishop Gordon's aliases: he was Bishop of Nicopolis.

[4] 'I should be very glad to see my old Nest, for which I still retain a very warm attachment; but fear it will not be in my power'. SCA, BL, George James Gordon to John Geddes, 19 June 1765. George James Gordon was a student at Scalan from 1717 to 1725. He was ordained there in September 1725 and came back as master in 1727, an office he fulfilled for over ten years.

[5] George James Gordon was ordained alongside Hugh MacDonald, the first Vicar Apostolic of the Highland District. Francis MacDonald was ordained in 1736, but apostatised a few years later to become a Royal Bounty catechist. John Gordon was ordained in 1754 and was appointed to the mission of Glenlivet, but died in 1757, aged 28.

[6] More often than not, that meeting was held in the summer at Scalan and was the occasion for drafting 'Italian' letters, that is to say letters to various dignitaries of the Roman Catholic Church in Rome.

[7] See note 4 above. Both the aliases and the metaphors were too transparent by then to afford the people involved any protection.

through the regular post. While John Geddes was at Scalan, his letters went through Aberdeen before being put in the 'south post' by George James Gordon. On at least one occasion, Gordon urged Smith not to forward a letter Geddes had written in February 1763 to Cardinal Spinelli, which contained details about two boys at Scalan that he hoped to send to the Scots College in Rome. What Gordon resented was Geddes's determination to have as many boys as possible instructed abroad, without attempting to see them through their studies at Scalan. That this was a live issue throughout his mastership, is manifest from Geddes's answer to a letter Bishop Smith had written to him in 1765, making it clear to the master what his design for Scalan was. Geddes first reassured his Bishop that he would 'endeavour to follow [his] plans to the utmost of [his] Powers' but then concluded: 'I also hope, it will be easy, if it shall please God to continue the present Peace we enjoy, both to prepare here some Apprentices for foreign Shops, which is certainly a thing of the utmost Consequence, and also to bring others farther on in their Business, according to your Desire'.[8]

As a matter of fact, in the eighty-odd years of its existence, Scalan was mostly used as a junior seminary, which did not make it the less essential to the Mission, as it supplied the Scots Colleges with students and, by way of consequence, the Mission with priests.

However, Scottish Bishops seemed to be of two minds about Scalan. They were in a constant dilemma—on the one hand, they were alive to the pivotal role it fulfilled; on the other, they were but too conscious that it was a drain on the Mission's scant financial and human resources. Besides the cost of maintaining boys there, often for as long as five years, Scalan deprived the Mission, as the Bishops saw it, of an able 'hand'—the master could have been otherwise employed, at a time when the Catholic Church in Scotland suffered from an endemic shortage of priests. This dilemma was particularly acute when Bishops had to make choice of a master. Their correspondence amply testifies to the fact that they had no doubt that the future prosperity of the Mission was mainly contingent on the success of the Scalan master, but they were painfully aware that supplying the seminary with a competent master put the present standing of the Church at risk by leaving whole areas unattended by a resident priest. In the last resort, Bishops seem always to have favoured the needs of their existing flocks over those of subsequent generations, though after much agonizing. In 1774, for instance, at their annual meeting, the Bishops for both districts refused to grant John Geddes the second master he had been craving for the Scots College at Valladolid. In the letter he wrote him, Bishop Hay gave the rationale for this decision:

> The Souls of our present flock are our immediate, our first & principal Charge; we must render an account of them to almighty God, but not of those who may come after us; and tho it be our Duty to take all reasonable care to provide for futurity; yet when this cannot be

[8] SCA, BL, John Geddes to Alexander Smith, 3 March 1765. The 'peace' refers to a lull in the priest-hunt in the north-east, but also to the end of the Seven Years' War, as the naval blockade made it extremely difficult for Scottish boys to get to the Scots Colleges abroad.

done without the manifest ruin of those souls at present under our Charge, it is evident that Duty in this case obliges us to take care of the present & leave futurity to the Protection of Heaven.'[9]

What happened in the 1760s is a case in point. As early as 1761, the Lowland Bishops had pitched upon John Geddes, in charge of the mission of Shenval since his return from the Scots College in Rome two years earlier, as the perfect person to become the master of the seminary. However, they could find no replacement for him and had to put their scheme in abeyance until August 1762, when William Duthie eventually accepted to take over the mission of Shenval.[10] Geddes had barely been two years at Scalan, when pressing demands were made on the Bishops to send him to the Scots College at Douai. Though they refused to remove Geddes to the Continent in 1765, heaping encomiums on his skills as master, he was ordered to leave Scalan in 1767 to replace George Hay at Preshome, an essential mission in the Enzie, so that, when push came to shove, even the man, whom Bishop Grant claimed in 1765 was 'so very necessary' at Scalan, was allowed to be replaced by John Thomson, a relatively indifferent master,[11] who was in his turn succeeded by John Paterson, a former pupil of John Geddes's there. Paterson, who self-confessedly wished 'to be stil'd a second Mr John Geddes',[12] remained in office until his untimely death in 1783.

In between them, Geddes and Paterson trained the bulk of the priests who served the Lowland District into the nineteenth century.[13] Though a sizeable part of eighteenth-century Scottish Catholic correspondence is no longer extant, what has survived bears testimony to abundant epistolary links. The Scalan masters exchanged letters with their bishops and senior priests on the one hand, and with their former pupils on the other. Both Geddes and Paterson were compulsive readers and writers of letters, but while the former had a very extensive European network both within and outwith the Catholic community, Paterson's was extremely circumscribed. There is no trace remaining of his passive correspondence, that is to say of the letters he received, except for three 'collective' letters Bishop Hay wrote between September

[9] SCA, BL, George Hay to John Geddes, 9 March 1774.

[10] The Bishops' plan had foundered the previous year on another priest's categorical refusal to move to Shenval.

[11] George Hay had been appointed procurator, which meant he had to be stationed in Edinburgh. Geddes himself later confided to Hay: 'I own to you, that I thought nothing could excuse the sending Mr Thomson to Scalan […] but downright Necessity'. SCA, BL, John Geddes to George Hay, 28 November 1769.

[12] Royal Scots College, Valladolid (now Salamanca) [thereafter RSC], John Paterson to John Geddes, 9 June 1776: 'My ambition is, to be stil'd a second Mr John Geddes, this I hope you will say is not unlawful.'

[13] Alexander Cameron, James Cameron, John Farquharson, John Gordon (later vice-rector at Valladolid), William Hay, Alexander Innes and John Paterson were taught by Geddes. Thomas Bagnall, Andrew Dawson, Alexander Farquharson, John Gordon (John Geddes's nephew), Peter Hay, Lachlan MacIntosh and George Mathison by Paterson. The last two mentioned only spent a few months in Scalan under Paterson's supervision, but left the seminary to be trained by Geddes at Valladolid. One further priest, Paul Macpherson, must be added to the list. Though he was a pupil of John Thomson's at Scalan, he was first taught reading and writing by Geddes, his local priest, and later studied for the priesthood at Valladolid while Geddes was Rector of the College.

and November 1772.[14] Considering Paterson's declared passion for letters, which sometimes seemed to verge on fetishism, it is highly improbable that he did not keep any of them and it can only be surmised that his passive correspondence was either destroyed or lost after his death. His extant letters, mostly addressed to John Geddes and Bishop Hay, cover the years of his mastership (1770–1783). John Geddes's passive and active correspondences have fared better, though a substantial amount of letters is no longer available. While he was master, Geddes's letters had the same outlook as Paterson's: they were mainly to and from the Lowland bishops, Alexander Smith and James Grant, and George James Gordon, 'Scalanensis'. Both Geddes and Paterson, then, had privileged epistolary links with one of their predecessors at Scalan. It is worth noting that, on both occasions, it was at the Bishops' instigation that the masters entered into correspondence. When Bishop Grant entreated John Geddes, then in Spain, to write to John Paterson who had just been appointed to the seminary at Scalan, he reminded him of what his own situation had been:

> I would Likewise wish, you would take the trouble to write to Mr Paterson. [Y]ou will have readily heard that he has been lately settled in Scalan. [H]e is quite sensible how far he is from having the qualifications of most of his predecessors placed there by our most worthy superiors, who were so happy as to have a great number of choices before them, whereas at present Settlements of this kind must rather be made by necessity than choice; but in the main he is a good Solid judicious Lad, extremely willing to do well and to listen to proper advice: when you was placed where Mr Paterson now is, you will remember that the late worthy Mr George Gordon, who was very capable of it, wrote many a long letter to you relative to your conduct in that important Station. I beg therefore that you in your turn will be so good as write to honest John your late pupil on the same subject. I'm fully convinced that he will shew all due regard to your advice, which your experience acquired by being some time there will render you more capable to give him than many others can be supposed to be.[15]

In that regard, the letters John Geddes and John Paterson wrote each other are of particular interest, as Paterson was not only Geddes's successor at one remove, but also, as Grant himself points out, a former pupil of his, and their multilayered relationship often shows through in Paterson's letters. The contents of those penned by Geddes can only be conjectured from his former pupil's, since none of them is extant. Once he left Scalan, Geddes kept in regular—though not frequent—contact with most of his pupils, mainly once they had themselves left Scalan to study at one of the Colleges on the Continent. He may have made the same recommendations to them, as he did to George Mathison, his pupil at Valladolid, who had just started officiating as priest in Scotland:

> I will be glad to hear from you once every three Months, or at least once every half year. I would wish that you and some others of my Acquaintances would let one another know

[14] Besides Paterson, the recipients of these letters were John Thomson, one-time master of Scalan and by then in charge of the mission of Strathavon, and the newly-ordained Alexander Cameron, who was stationed at Tomintoul.

[15] RSC, James Grant to John Geddes, 28 September 1770. Soon after he arrived at Scalan, Geddes wrote to Bishop Smith: 'I am very sensible of the Advantage I may draw from corresponding with Mr George at Abd^{n}, & shall most willingly obey you in having Recourse to him in Doubts'. SCA, BL, John Geddes to Alexander Smith, 27 November 1762.

> when you favour me with a Letter, that you may not write me the same Pieces of News, and that your Letters at a due Distance may give me the satisfaction of knowing how you do, and what is going on in our Country.[16]

The major difference between Paterson and Geddes is that the former's mastership almost encapsulates his whole adult life, while the latter lived on to become Rector of the Scots College at Valladolid and eventually Coadjutor to the Lowland Vicar Apostolic. This is reflected in their surviving correspondence. Letters to and from Geddes run the whole gamut of epistolary exchanges—from a young master seeking and receiving advice to a Bishop who, when needed, brought his authority to bear on the Lowland priests, his former pupils included. In between, Geddes was a highly respected Rector, whose former pupils turned to for advice, while he himself, beset with doubts when apparently promising students 'miscarried', sought reassurance from his Bishops. Paterson received letters from those of his pupils who had gone on to study at Valladolid, though it would seem he seldom wrote to them.[17] The epistolary links between Paterson and his former pupils must of course have been of a somewhat different nature from those that existed between his predecessor and his own pupils. Paterson was too well aware that the seminarists at Valladolid would welcome news from home but, with Geddes on hand, stood in no need of guidance. Conversely, Geddes's pupils who trained for the priesthood at the Scots Colleges of Rome or Douai were eager to have his views on what they ought to read and study, and their former master was only too happy to oblige. On one occasion, Geddes's letter did not reach Paul Macpherson, then at the Scots College in Rome, who begged him to reiterate what he had first written, which he did, to Macpherson's obvious satisfaction:

> If my gratitude towards you could be increased, I would have great motives to do so for … the prudent advices concerning my studies which you was pleased to write me …. I promise you I will do all my endeavours to observe your wise counsels, and I am sure my companions will do the same. I am only sorry that I know but very little of the Greek language, which is one of your prerequisits.[18]

---

[16] SCA, BL, John Geddes to George Mathison, 3 May 1779. Obviously, Geddes would have wanted his Scalan pupils to let him know about the situation in the Scots Colleges they studied at, which they did. Indeed, the letters Geddes received from the Scots College at Rome in the late 1760s and 1770s were the Bishops' main source of information on that College's dysfunctional state.

[17] References to such letters can be found in Paterson's correspondence with Geddes, but none of them is extant. See, for instance, RSC, John Paterson to John Geddes, 6 July 1774: 'Give my kindest Compl[ts] and blessing to your young folks, especially to my old Pupils[.] I would willingly write to them, but really time will not permit, hope therefore that they will excuse me at present, next time I write to you by post, shall add a few lines in that letter to them…I received Jo: Gordon & George Mathison's Letter do thank them for their kind rembrance [sic] of me, shall write them as I said before some other time.'

[18] SRC, Paul Macpherson to John Geddes, 4 July 1776.

That this was not a one-off is manifest from Macpherson's regular expressions of gratitude, for instance in April 1777, just months before he left Rome for health reasons and went to complete his training at Valladolid under Geddes's direct supervision: 'I return you many thanks for the advice you was kind enough to give me concerning Controversy and speculative Divinity. I shall expect with anxiety something about Morality whenever you will find it convenient.'[19] That Geddes was wont to offer such help is clear from a letter John Farquharson, another of his pupils, then at Douai, wrote to him:

> In case you'll think proper to honour me with a line, I hope in going on, in your ordinary advertising Strain, you'll not faill to point out to me what books of Divinity are the most fit? what points I ought to apply myself to, in a particular manner &c &c. I may tell you, that as yet I'm an entire stranger to Moral Divinity, (which I fancy to be the most necessary for the Mission)...[20]

On the face of it, religion cut a surprisingly small figure in the correspondence of Scalan masters with their superiors which is mainly taken up with 'business', that is to say the daily management of the seminary. This included securing the tack for the farm, hiring and sacking servants, providing foodstuffs on the best possible terms, as well as making and keeping the building fit for its purpose and, most important of all, balancing the books.[21] This was no mean task for a single man, who was also in charge of the education and spiritual direction of, at times, as many as a dozen boys. Financial matters loom large in the letters written by John Paterson. When he took over from John Thomson, he found Scalan drowned in debts incurred by his predecessor, who had overreached himself in his attempt to improve the building. Though this was to be a major source of 'floughtiness' in Paterson's early days as master, he got the better of his anxiety and, towards the end of a long letter to Bishop Hay, wrote, tongue-in-cheek: 'I suppose you are wondering against this time that I am not speaking about money for this place'.[22] The Mission was not only in straitened circumstances, it was also, as was hinted earlier on, short of priests. One way of getting round the difficulty of providing for Scalan was to have its master officiating as priest, answering calls from Catholics living in nearby missions that were vacant, such as Shenval or Strathavon. In the very first letter he wrote to Bishop Smith on settling at Scalan, Geddes adverted

---

[19] SRC, Paul Macpherson to John Geddes, 24 April 1777.

[20] SRC, John Farquharson to John Geddes, 8 February 1776.

[21] When Geddes took up his position in 1762, Scalan appeared to be lacking even in the bare essentials: 'I'm sorry your house is so ill provided of accommodation and the most necessary implements. But tho' the circumstances of times do not allow to be at great expences in the present uncertainties, yet I think that, at all events, the most requisite things for a decent tables [sic], clean beds &c should be got. Sure I am Mr Rob^{n} will agree to this and allow the expences, if you represent the matter properly to him.' SCA, BL, George James Gordon to John Geddes, 9 June 1763.

[22] SCA, BL, John Paterson to George Hay, 20 October 1771.

to what he foresaw would be a major hurdle: 'I find I will have a good deal of Difficulty arising from the Two different Employments I will have but I shall do the best I can'.[23]

In spite of his repeated absences, the master was at the head of a closely knit community. In letters, the boys were referred to as the master's 'family', and Paterson wrote of his 'Bairns'. Considering the fact that most boys entered Scalan in their early teens—some were as young as eight years old—and spent a number of years there, it comes as no surprise that they would endow the master with a paternal role, which outlasted their time at Scalan. In 1776, 'after a Silence of Six whole years', John Farquharson wrote a long letter to John Geddes, looking both backward and forward:

> I confess indeed, it's a mystery for me altogether unfathomable what could have been the motives which have induced you to testify so much regard, such tender and Cordial affection towards one, who did not in the least deserve it. What grieves me is my not corresponding & complying, to such fatherly care. All that I can say, is, that you have not bestowed your affection on an ungrateful heart. Often do I recall with pleasure the short but agreeable years, which I passed under your care, on the Banks of Crombie. When I reflect on my returning to Scotland, I'm already somewhat sorry, to think that I shall not have the Satisfaction of having you near me, in order to consult you, and get your salutary advices: After all, I do not as yet altogether dispair, of meeting with you in Scotland, or at least of passing some part of my life, under your paternal inspection.[24]

That may have been a hint at what a number of people expected, and Farquharson hoped, that is to say, that John Geddes would become Bishop Hay's Coadjutor on Bishop Grant's death, but it was also very much the expression of his feelings towards his former master: years before Geddes became Bishop, Farquharson would address him as his 'Dear Father' and subscribe himself his 'respectful & obedient child' or 'most Dutiful & respectful Son'. Even though Paterson undoubtedly looked upon his predecessor as the epitome of the good master, on a number of occasions he suggested that they were now colleagues of a sort, sharing in the responsibility to train priests for the Mission, and, more often than not, addressed him as his 'Dear Friend'. On the other hand, it is worth noting that Paterson did use similar formulæ in his letters to Bishop Hay. After his spell as a pupil at Scalan, Paterson had stayed many months under Hay's supervision in Edinburgh getting ready for his ordination, and as a result he looked on the Bishop as a 'loving',

[23] SCA, BL, John Geddes to Alexander Smith, 13 September 1762. Six years later, looking back on his tenure as master with mixed feelings, lamenting the 'money & Pains…lost on Boys of whom we can have little Hopes of their ever being of Use', he reminded Bishop Grant of the extra burden he had had to shoulder: 'I must also take Notice, that during almost all the time I had the Care of that House, I had it not in my Power to give one half of the Attendance I saw necessary; so that I thought it but Justice to attribute a great deal of the things, that did not please me to the Want of a prudent Superiour present with them more constantly'. SCA, BL, John Geddes to James Grant, 7 March 1768.

[24] SRC, John Farquharson to John Geddes, 8 February 1776. Scalan lay by the banks of Crombie Burn.

'affectionate' Father.[25] Hay reciprocated the feeling, observing in his first common letter to three young priests that he looked upon Paterson as 'a Child of [his] own'.[26] Alexander Cameron was one of the addressees of that letter and a former pupil of John Geddes's too. A poor correspondent,[27] he concluded the first in a series of letters acknowledging his deficiency pledging to mend his ways:

> as I cannot expect you should write me, without my writting to you: let bygones be bygones, and, for the future, I promise, you shall find me a better child. It will give me the greatest pleasure to see you write, with the same freedom, I flatter myself, you would speake to me, on any subject whatsoever. I have the vanity to think you will find me as candid and ingenuous — as others.[28]

He thus emphasized the role of correspondence as an intrinsic part of the relationship between master and pupil, as well as the similarity, as he saw it, between written and oral communication. When Cameron took over from Geddes at Valladolid in 1780, he went so far as to write to his predecessor, who had just left Spain after spending many a month putting his former pupil through his paces: 'One can write many things they could not well say: whilst you were here we never spoke of some things which might have been discussed, perhaps not altogether uselessly.'[29] That letters were 'written conversations between friends', substituting for the physical presence of correspondents, was a commonplace, but Cameron's views were exceptional—more often than not, correspondence was perceived as a second-best choice.[30] I have written elsewhere of the ambivalent feelings of writers deprived of the possibility to meet face to face and the various ways in which they attempted to make the absent present.[31] Many correspondents touched on their inability to set down in writing their innermost sentiments and thoughts, often in the course of attempting to do so. In some cases, though by no means all, the priest's uneasiness was partly due to the clandestine status of the Catholic Church in Scotland—it was felt that some things were best left unwritten lest they were used against the whole Catholic community if the letters were intercepted by malevolent persons. The inadequacy of correspondence was resentfully noticed, even—and mostly—by such a compulsive letter-writer as Paterson. He regularly wrote long letters to George Hay, giving thorough accounts of what passed at Scalan. However, he longed for an interview

---

[25] See, for instance, SCA, BL, John Paterson to George Hay, 25 December 1770: 'Much Honour'd Sir & Loving Father'; John Paterson to George Hay, 20 December 1771: 'I am Much Honour'd Sir, and most affectionate Father, your most loving and most obed^{t} Child while I am John Paterson'.

[26] SCA, BL, George Hay to Alexander Cameron, John Paterson and John Thomson, 12 September 1772.

[27] SCA, BL, Alexander Cameron to John Geddes, 25 July 1783: 'I find, amongst my other papers, a more than half-finished letter to you, in which I tell you I had the pleasure of your letter of 26 Septr by last post'.

[28] RSC, Alexander Cameron to John Geddes, 9 June 1776.

[29] SCA, BL, Alexander Cameron to John Geddes, 26 March 1781.

[30] SCA, BL, Paul Macpherson to John Geddes, 28 October 1783: 'as I cannot expect to have the pleasure of seeing you soon I must write you some thing in regard of myself'.

[31] Prunier 2007.

with his Bishop: 'If I were at present just hard by you most willingly could I open the state of my whole conscience to you. I hope I will see you here this summer'.[32] He soon after reiterated his desire to spend some time with Hay at Scalan in a missive which deserves to be quoted at length:

> I received your last of date 13th Ult. and read it you may believe me with the greatest of pleasure, because I saw it had come from a heart most desirous of my welfare, and that the words of it had been dictated by the H. Ghost himself; wherefore with the grace & assistance of this same divine Spirit, I shall put in practice what you demand of me for my good to do. These H: S. are strong words, The great accounting day, arise ye dead and come to judgment, but which alas! have hitherto sounded but very little in my ears, God grant that I may now begin, and begin in earnest; all that I can say for myself, at present, is, that I really feel in myself an earnest desire of becoming good, and of serving that Adorable Being faithfully, who has already bestow'd upon me so many and so great favours. O Sir pray, and pray earnestly for me, that God Al: may give me constancy and perseverance in the doing of good, for this is my great, and very great fault, that I speculize much, but practise little; that I build many & very many Castles in the air, but never, nor never come to any kind of Maturity; that I know many times what should be done, but often, & very often fail much in the performance; and he who knows his Master's will, but does it not; &c O may God Al: grant that while I am preaching to others, I myself may not become a Cast away. When you come here, which [I] hope, & earnestly wish will be this summer, I shall lay open to you the whole state of my Soul and let you form a judgment, and tell me what you think. Your directions, and rules, I intend above all to follow; so correct, and chastise, and amend, whatever you see wrong, and I shall become as wax, to go into any form you please. This your last Letter to me so much pleas'd Mr Guthrie that he must have a Copy of it, and the original I am to keep, and to keep for myself alone.[33]

Some *raisons d'être* of correspondence are adumbrated in Paterson's letter. Though it could not wholly replace a conversation, it still could fulfill a number of essential purposes, in particular as far as Scalan masters and their pupils were concerned. Letters were a channel of advice, not only from the master to his former pupils, as so often happened in Geddes's correspondence, but also from senior clergy to the master himself. It could come in various forms: in his early days as master, Paterson thanked Geddes for his 'good & friendly advice'[34] while he solicited Hay's 'correction, and charitable advice', which was apparently forthcoming and seems to have been of a more peremptory nature, verging on demand. Hay certainly felt it was his duty to provide young priests with epistolary instruction, as is abundantly clear from the three letters he wrote for the benefit of John Paterson, John Thomson, and Alexander Cameron in 1772.[35] In the first letter, Hay recounted the 'means in [his] power to contribute to give [them] all possible assistance', and suggested the three of them should hold 'frequent Conferences' and give him a detailed account of such meetings. He then went on to the means accessible to him:

> I should communicate to you in writing from time to time what necessary advices for your Conduct may occur to me either from my own small experience & observation, or from

[32] SCA, BL, John Paterson to George Hay, 22 March 1771.

[33] SCA, BL, John Paterson to George Hay, 6 May 1771.

[34] RSC, John Paterson to John Geddes, 13 November 1770.

[35] The three priests were 'neighbours'. Only three letters are extant, but there might have been more.

> what lights I may have got from others. I am sorry that the many avocations & Employments I have got must hinder me from doing this so much or so frequently as I could wish yet as I look upon this as of no less importance than many other things I have to do I shall not fail, with God's assistance, to write to you in common as often as I conveniently can.[36]

Hay thus highlighted the crucial role of letter-writing in making sure the Mission's priests, the Scalan master included, were not only instructed, but also suitably supported and encouraged. His statement that any future letters would be written to the three priests jointly indicates that he saw these very much as pastoral letters whose content was destined to be shared. That this was also often the case in practice with letters addressed to individual priests is corroborated by recurring references to letters being read to, or by, other priests. However, as the quotation from Paterson emphatically demonstrates, a letter could not be reduced to its content: while the words were readily shared and copied, Paterson was not to part with the original epistle. There are obvious practical reasons for Paterson's unwillingness to relinquish Hay's letter. Still, the extant correspondence testifies to the fact that letters, regardless of their actual content, were considered as the embodiment of the relationship that obtained between correspondents—what bonded them was their Catholic faith and more particularly their vocation. Indeed, my contention is that there is much more religion in the correspondence of the Scalan masters than meets the eye at first. William Hay, who had been John Geddes's pupil at Scalan, was eventually dismissed from the Mission in 1783 for inappropriate behaviour. In the months leading up to his removal, he had regular, though tense, epistolary exchanges with his former master, now Coadjutor to the Vicar Apostolic for the Lowland District. At the height of the crisis, William Hay, who never alluded to his Scalan days,[37] wrote to Geddes: 'the very writing me, makes me think that you pray for me', pointing to the fact that letter-writing was an eminently religious act.

If correspondence was a 'written conversation', it also assumed a cathartic function, sometimes almost akin to a substitute for auricular confession. On two occasions at least, Paterson wrote to Geddes of his ability to 'unbosom' himself to his former master. In his letter dated 28 December 1782, it came about as he recalled their 'last tête à tête over Cairndoulack',[38] but in the second instance, a month later, it clearly referred to correspondence:

> I shall not be one Letter in your debt. Nay, I shall think myself honestly deal'd with, and, truly, honour'd, if you shall be pleas'd to accept of two letters from me, for one from you. I hope you will not take it amiss that I have propos'd this bargain to you, tho' bargains be not properly the work of the Clergy. I ever found it a happiness, for myself, when I met with a person, to whom, I thought, I freely could unbosom myself. It delights then to clatter out, the Childish thoughts themselves; because one is not afraid of their being

[36] SCA, BL, George Hay to Alexander Cameron, John Paterson and John Thomson, 12 September 1772.

[37] It is difficult to form a judgment from William Hay's letters whether Geddes did—the Coadjutor's letters are not extant.

[38] SCA, BL, John Paterson to John Geddes, 28 December 1782.

> ill interpreted.— How often have I plagu'd you, with heaps of buff, and Stuff, and as often did you Condescend to my weakness and chas'd away my fears; and no Body ever hear'd of it.[39]

Paterson, for one, seems to have been convinced that letter-writing in itself was somehow performative. A few months after he settled at Scalan, he lamented that he had 'so much fallen back' since he had left Hay at Edinburgh and that 'Meditations, & Spiritual Book ha[d] been entirely forgotten'. He then confided to Hay : 'you see how soon a person's fervour will Cool, however I hope this very telling of you will be a good means to wear the temptation off'.[40] It could then be argued that, irrespective of their contents, letters were essentially religious, not only in their inspiration,[41] but also in their effect. '[Y]our words have a wonderful effect in me the reason of which I cannot easily account for', Paul Macpherson noted in one of his early letters to Geddes.[42] When John Thomson was sent to Rome as the Scots Agent, he entreated his predecessor at Scalan to correspond with him: 'your letters will be a Cordial to me in my present Solitude for I assure you I have few friends here & many Enemies. My fortitude has been put to the trial … but with God's help I shall persevere.'[43] As for John Paterson, he craved Hay's letters, including those that were 'somewhat sharp':

> I beg of you for the future, still continue to correct me in whatever way you please, which shall be always agreeable to me, & with God Almighty's help I shall endeavour to do better; but this also I must beg of you, when you write me something severe, that you would soon after send me a Letter full of comfort, for I have need of that from time to time, at any rate.[44]

In the context of the clandestine Scottish Mission, allaying the fears of the Scalan master, and bringing him some reassurance and encouragement, ran deeper than personal comfort. In an oft-quoted passage of his 'Brief Historical Account of the Seminary of Scalan' penned in June 1777 for the benefit of his students at Valladolid, Geddes asserted that '[t]he time, by the Goodness of God, will come, when the Catholic Religion will again flourish in Scotland; and then, when Posterity shall enquire, with a laudable Curiosity, by what means any Sparks of the true Faith were preserved in these dismal Times of Darkness and Error, Scalan and these our Colleges will be mentioned with Veneration.' He went on to praise 'those Champions, who stood up for the Cause of God, and far from being carried away by the Torrent of Prejudice, in spite of all Opposition and Delusion, went on stedfastly in their

[39] SCA, BL, John Paterson to John Geddes, 31 January 1783.

[40] SCA, BL, John Paterson to George Hay, 11 March 1771.

[41] See, for instance, SCA, BL, John Paterson to George Hay, 6 May 1771, quoted above.

[42] RSC, Paul Macpherson to John Geddes, 5 April 1776.

[43] SCA, BL, John Thomson to John Geddes, 16 February 1783. When Geddes left Spain, John Gordon, his former pupil at Scalan, then Vice Rector of the Scots College in Valladolid, made a similar request. Though he was in much friendlier surroundings, he used the same words: 'Your letters will be at all times a cordial to me'. SCA, BL, John Gordon to John Geddes, 26 February 1781.

[44] SCA, BL, John Paterson to George Hay, 14 September 1770.

Duty to their Maker'.[45] The champions he had in mind were not only the priests who had received their early education at Scalan, but also their masters, whose task it was to ensure the Mission was supplied with clergymen in the future. Theirs was a heavy responsibility. Sometimes it seemed almost impossible that they should grapple single-handedly with all their duties and they were bound to be oppressed by doubts as a result. Even Geddes at some point contemplated leaving Scalan to return to Shenval.[46] As for Paterson, he particularly prized Hay's letters, to whom he wrote, begging for a continuation of their correspondence: 'every time I receive a Letter from you it gives me new vigour'.[47]

The vigour Paterson found in reading Hay's letters was the vigour to go about his daily business, but it was also the vigour to persevere in the face of adversity, the vigour to put Scalan to good use and eventually the vigour to keep up the Catholic faith in Scotland, so that if, as Geddes maintained, Catholicism partly owes its survival in Scotland to Scalan, it surely owes as much to the correspondence of the Scalan masters.

References

Johnson, Christine. 1983. *Developments in the Roman Catholic Church in Scotland 1789–1829*. Edinburgh: John Donald.

McMillan, James F. 1999. Mission accomplished? The Catholic underground. In *Eighteenth-century Scotland: New perspectives*, ed. T.M. Devine and J.R. Young, 90–105. East Linton: Tuckwell Press.

Prunier, Clotilde. 2004. *Anti-catholic strategies in eighteenth-century Scotland*. Frankfurt am Main: Peter Lang.

Prunier, Clotilde. 2007. Monseigneur(s) et ami(s): la correspondance de John Geddes et de George Hay, 1759–99. In *La lettre et l'intime. L'émergence d'une expression du for intérieur dans les correspondances privées (17e–19e siècles)*, ed. Paul Servais and Van Laurence Ypersele, 55–79. Louvain-la-Neuve: Bruylant-Academia.

Szechi, Daniel. 1996. Defending the true faith: Kirk, state and catholic missioners in Scotland 1653–1755. *The Catholic Historical Review* 82: 397–411.

Watts, John. 1999. *Scalan, the forbidden college 1716–1799*. East Linton: Tuckwell Press.

---

[45] RSC, Geddes Papers, John Geddes, 'A Brief Historical Account of the Seminary of Scalan'. The text of another copy, in the Scottish Catholic Archives, can be found in 'The College for the Lowland District of Scotland at Scalan and Aquhorties: Registers and Documents', W. J. Anderson (ed), *Innes Review* 14 (1963), 89–212.

[46] See SCA, BL, George James Gordon to John Geddes, 19 June 1765: 'I do not think you should entertain any thoughts of exchanging your present business, which plainly appears to be the place in which God would have you to be, for your former or any other station. We are allwise best and may expect the greater blessing and happy success in the situation the Divine Providence puts us in, by the order of good and wise Sup^{rs}.'

[47] SCA, BL, John Paterson to George Hay, 12 November 1771. It is not clear why Paterson apprehended their correspondence would lapse at that point, except for the fact that he was fully settled at Scalan by then and might have feared that Hay would no longer feel it necessary to write regularly to him.

Part III
Religion, Science and Philosophy

Chapter 9
Utopian Intelligences: Scientific Correspondence and Christian Virtuosos

Claire Preston

> *The elements of human knowledge, that is to say, observations, experiments, mechanical inventions, medical and chemical secrets ... are scattered and dispersed throughout the globe. I think there is no better way of assembling them than by constantly bringing together the zealous and the inventive in regular correspondence.*[1]
>
> (Henry Oldenburg to Stanislaw Lubienietzki)

A year before his death, Robert Boyle published *The Christian Virtuoso* (1690), one of a number of works in which he discussed the physico-theological association of scientific practice and religious understanding. 'There is no Inconsistence,' he announces at the outset, 'between a Man's being an Industrious *Virtuoso*, and a Good *Christian*…'[2]

> [T]he Fabrick and Conduct of the Universe ... very much indisposeth the mind to ascribe such admirable Effects to so incompetent and pitiful a Cause as Blind Chance, or the tumultuous Justlings of Atomical Portions of senseless Matter; and it leads directly to the acknowledgement and adoration of a most Intelligent, Powerful and Benign Author of things…[3]

In addition to his experimental philosophy and his theoretical speculations on natural phenomena, primarily chemistry and physics, he was also a vigorous promoter of Christian doctrine, through his moral works and in his evangelical support

---

[1] 23 July 1666 (*The Correspondence of Henry Oldenburg*, eds A Rupert Hall and Marie Boas Hall, 9 vols (Madison: University of Wisconsin Press, 1965–73) III, 192, trans from Latin by eds (thereafter *OC* with volume and page-number).

[2] *The Christian Virtuoso* (1690), A[1r].

[3] *The Christian Virtuoso*, 9.

C. Preston (✉)
Department of English, University of Birmingham, B15 2TT Edgbaston, UK
e-mail: c.e.preston@bham.ac.uk

A. Dunan-Page and C. Prunier (eds.), *Debating the Faith: Religion and Letter Writing in Great Britain, 1550-1800*, International Archives of the History of Ideas 209,
DOI 10.1007/978-94-007-5216-0_9,

of Christian missions, especially to the Arab world.[4] If the finality of culminating works is significant, it is telling that Boyle's last foray into print was the clearest of his many attempts to yoke these two halves of his intellectual life, natural philosophy and Christian theology, to show that the study of nature conduces to, rather than competes with, belief. Like most of his scientific contemporaries, however, he subscribed equally to the policy of the Royal Society and its precursors in England of specifically *excluding* religion from its discussions. This was a necessity during the civil wars and the Interregnum, when natural philosophers on opposite sides of the religious-political divide in England had to agree to discuss science in a neutral zone: scientific investigation requires stability, and successful investigation required freedom from sectarian tincture.

The embargo on discussion and reference to faith did not, however, obviate the overarching theological framework within which most natural philosophers located their investigative programmes. The natural theology of the mid- and later-seventeenth century explicitly intepreted the phenomena of the natural world—their beauty, order, consistency, and pattern—as divine signatures or deliberately placed tidings of God's omnipotent plan. The scientists regarded their work as godly, and the tools that allowed them to forward their scientific programme, morally virtuous. One of those tools was correspondence, a necessary activity for very widely dispersed participants (in many parts of the British Isles and throughout the Continent as far east as Constantinople); but it was posed as an ideal cooperative and collaborative community, righteous, utopian, edenic, located in the Baconian task, in the diminished post-lapsarian world, of re-establishing true knowledge through the advancement of learning. Although they explicitly forbade the *discussion* of faith, at the same time Christian virtuosos made their letter-networks part of God's work. It is useful, therefore, to acknowledge in any discussion of epistolarity and faith that spiritual belief could be expressed through correspondence in covert ways, as actions rather than as utterances, and as the substructural *donnée* of intellectual undertakings. This discussion of scientific correspondence is therefore not primarily about belief; instead, it suggests that even the apparently theologically neutral writings of the corresponding scientists were profound articulations of faith, that their rhetorical construction and maintenance of a virtual, utopian community through correspondence carried out the primary duties of the Christian virtuoso: to magnify God in investigating the creation; to operate in a highly conscious mode of collaborative intellectual generosity, especially across religious, national, and political divisions, as a form of Christian *caritas*; and to sustain this selfless intellectual nation in order to recreate an undamaged, or less damaged, world of eirenic, godly enterprise for the benefit of humankind.

In 1647, the Anglican Boyle was discussing Christian utopian schemes—Johannes Andreae's *Christianopolis* (1619), Tomaso Campanella's *Civitas Solis* (1623), and John Hall's work-in-progress, *Lucenia*—with Samuel Hartlib, the Protestant eirenicist, pansophist, and intelligencer, and the publisher of Gabriel

[4] Boyle sponsored the Arabic translations of Christian apologetic works to be distributed by the Levant Company.

Plattes's *Description of the Famous Kingdom of Macaria* (1641). The scientific correspondence network (Kronick 2001, 32)[5] of which they were both part was clearly another such commonwealth in Boyle's mind: he refers to word of new scientific work as 'utopian intelligences', and he specifies the rhetorical conventions of such a commonwealth when he instructs Hartlib to omit any merely ceremonious phrases as 'but the froth (not to say the scum) of civility'. In 1690 John Ray disparaged poetic and oratorical ornament as concerned merely with words and not things.[6] Similarly, in 1666 Nathaniel Fairfax, a curious physician in Suffolk writing to Henry Oldenburg, excused his terseness as a specific against extravagant and copious writing: '...wherein Sr I have imployed thus many words & no more yt I may quoit ye paper into your hands, after ye bluntness of a Thorpsman, nor overmodishly wave it in my own, wth its answering honors & fottings, after ye antiknes of a kickshaw.'[7] And, recounting to Samuel Pepys a dream in which 'me-thought Mr Pepys and I were ... discoursing in his Library, about the Ceremonious Part of Conversation, and Visites of Forme, betweene well-bred Persons,' John Evelyn is all for ceremonial 'scum' and verbal 'kickshaws'.[8]

There is a clear disagreement here about rhetorical propriety: ceremonies and forms that Evelyn obviously regards as the height of intellectual ease and contentment in a civil face-to-face encounter are ones that Boyle and Fairfax seem to understand as a problematic rhetorical tendency in the writing of science, one that partly arises in the effort to distinguish the ideal, equal, and eirenic world of letters from the real, competitive, and troubled one of quotidian existence. To dispense with civil kickshaws as Boyle and Fairfax require was, presumably, to be more perspicuous, more plain, and more direct; in the contention between *res* and *verba* it was to give priority to writing about the things of nature (and thus of divine creation) over the verbal representation of social acts (and thus of the fallen world of human interaction). And yet this division is hardly stable even within the individual: Boyle would have argued that civility facilitates the ideal transmission and transaction of information and discussion; and even the kickshaw-hating Fairfax copiously deploys several striking tropes that seem to contradict his professions of plainness; he regretted

[5] I use the word 'network' to describe a variety of such informal and formally constituted correspondence groups, despite David A. Kronick's insistence that the term implies the formal conduct of communication by trade and finance organisations. On the nature of networks and 'circles', see also Judith Scherer Herz, 'Of circles, friendship, and the imperatives of literary history' in Claude J. Summers and Ted-Larry Pebworth (eds), *Literary Circles and Cultural Communities in Renaissance England* (Columbia and London: University of Missouri Press, 2000), 10–23, at 18–20; and Anna Nardo, 'A Space for Academic Recreation: Milton's Proposal and *The Reason of Church Government*', also in Summers and Pebworth, 128–47.

[6] John Ray to Tancred Robinson, 15 December 1690, quoted in Joseph M. Levine, 'Strife in the Republic of Letters' in Ann Bermingham and John Beaver (eds), *The Consumption of Culture 1600–1800: Image, Object, Text* (London: Routledge, 1995), 301–19, at 306n.

[7] *OC* III, 320, Nathaniel Fairfax to Oldenburg 25 Jan 1666/7.

[8] Evelyn to Pepys, 4 Oct 1689, in *Particular Friends: The Correspondence of Samuel Pepys and John Evelyn*, ed. Guy de la Bédoyère (Woodbridge: Boydell Press, 1997), 206 (hereafter Bédoyère).

that he lacked 'the knack of making flourishes wth my pen [or] I should have drawn ye picture of my thankfulness, in all its complementall ornaments, as farr as white & black would set off'.[9] If, as Gary Schneider has argued, the rhetorical patterns and habits of early-modern epistolarity evolved to make writing replicate orality and to make epistolary presence stand in for bodily, gestural presence, and the letter a form of physiognomy, a window on the soul (Schneider 2005, 109–42), these competing assumptions in Evelyn and in Boyle and Fairfax make plain that the erudite scientific letter needed to be situated in a precise but still uncodified socio-rhetorical terrain, and one that had philosophical-theological meaning. It was a terrain which was not identical to that of the private, social letter. In this essay I will argue that a Christian-utopian model of philosophical transaction required a distinctive set of rhetorical practices that at once functionally assisted and analogically enacted that work of reparation, that reconvening of the detritus of a prior, undamaged world being attempted by natural philosophy.[10]

If there existed an equivocal attitude to the special rhetorical nature and needs of epistolary science, it is in part because natural philosophers were at the same time engaged in a wider, ongoing debate about the value of ornamented language, tropes and figures, in discursive scientific writing—what Cowley hailed as Thomas Sprat's 'candid Style like a clear Stream ... comely Dress without the paint of Art'[11] even as he proposed his botanical poem *Plantarum* as gilded with 'the brightness of Stile'.[12] Any discussion of epistolary conduct among the scientists must therefore attend to their broader concerns about the linguistic 'etiquette of inquiry' (Biagioli 1996, 235)—to questions of what Boyle comprehended under the 'general name of *Style*': '*not only* the Language as it is concise or more Diffus'd, Embellish'd, or Unadorn'd, Plain or Figurative; *but also* under what Perspicuity, Veracity, Impartiality, Cautiousnes and other such Qualitys, have been aim'd at...'.[13] Boyle, as it happens, is an extremely useful index of how open the debate about scientific rhetoric was: throughout his career he was extensively thoughtful about the propriety and use of ornament and tropes in 'utopian' work, and did not scruple to employ divers generic forms; and his own equivocations on the subject of scientific language and linguistic ceremony are everywhere evident. He constantly introduced the very scum he disparaged, civil forms both in his epistolary and in his discursive scientific and philosophical works that could encourage, disagree,

[9] *OC* III 419–20, Nathaniel Fairfax to Oldenburg, 29 May, 1667.

[10] On the vision of the scientific project as 'a secular counterpart to millenarianism' and the reconvening of knowledge, see Michael Hunter, *John Aubrey and the Realm of Learning* (London: Duckworth, 1975), 19; and also my *Thomas Browne and the Writing of Early-Modern Science* (Cambridge: Cambridge University Press, 2005), 138–9.

[11] Abraham Cowley, 'To the Royal Society', stanza 9, *Verses Written on Several Occasions* in *The Works of Mr Abraham Cowley* (1688), 42.

[12] Cowley, *The Third Part of the Works of Mr Abraham Cowley, being His Six Books of Plants* (1689), b[1r]. .

[13] *OC* III, 162, Boyle to Oldenburg, 13 June 1666.

explain, clarify, congratulate, guess, insist, and deny while maintaining the canons of polite and cooperative intercourse.

Another key figure in this debate, and one especially concerned with the epistolary, was Henry Oldenburg—translator, advocate for the Royal Society, panegyrist for individual scientists, vanguard in establishing new contacts for himself and others, recorder of various announcements of discovery or invention, and by default the registrar, arbiter, and sometimes the deliberator of priority or truth.[14] He was as equivocal as Boyle about the rhetorical features of scientific writing, but as corresponding secretary of the Royal Society he necessarily insisted on civil froth and kickshaws, simply because they were the irreducible component in his task of soliciting and replying to correspondents, even to the extent, as we see below, of coaching various individuals in the proprieties of such exchange. But, he also maintained, 'verbal embellishment … I do not consider fitting for candid lovers of philosophy.'[15] In early-modern sciences, epistolary and discursive forms and tropes were geared to the scientific subject and to collaborative purpose. In particular, the standard civil observances of ordinary social epistolarity, observances that maintained and reinforced hierarchical order and emotional and spiritual relations were, in scientific correspondence, pointedly harnessed to the Christian/utopian project of collaborative investigation and discussion, so that conventional polite forms and ceremonies became the very fabric of the utopian. The rhetoric of scientific epistolarity in the service of scientific collaboration is a refinement and an enactment of the early-modern scientific project.[16]

According to Oldenburg, Boyle chose to cast his *Sceptical Chemist* as a dialogue because it represented *conversation*. This work, he explains, is a fiction wrought 'with great politeness, and with a propriety which may teach combatants how to oppose an opinion without wounding those who hold it'.[17] Interestingly, this definition textualizes the presence of speakers (Schneider 28–37)[18]; and it is a graceful definition, I think, of the kind of scientific civility Oldenburg so strenuously and self-consciously maintained in his own letters and refereed so assiduously among his many correspondents, a civility he pitched against the earlier scholastic tradition of disputation, an essentially *un*civil and *im*polite mode of discussion.[19]

[14] For example, between the French and the English over the first blood transfusion, or between Auzout and Hooke about priority in the invention of the filar micrometer (an instrument for measuring the angular distance between astronomical bodies). Oldenburg was in a sense the chronicler or historian of the virtual republic. Kronick (32–3) has described this as a 'gatekeeping' function, and one common to many of the 'corresponding' societies in Europe at this time in the persons of, for example, Mersenne, Peiresc, and Montmor.

[15] *OC* III, 192, Oldenburg to Lubienietzki 23 July 1666 (trans from Latin by editors).

[16] I have written elsewhere of non-epistolary rhetorical practice in early-modern science: 'English Scientific Prose: Bacon, Browne, Boyle' in Andrew Hadfield (ed.), *The Oxford Handbook to English Prose, 1500 to 1640* (Oxford: OUP, forthcoming, 2012).

[17] *OC* I, 422, Oldenburg to Christiaan Huygens, 7 September 1661.

[18] For textualisation, see Schneider, 28–37. See also James Daybell (ed.), *Early Modern Women's Letter Writing* (Basingstoke: Palgrave, 2001).

[19] *OC* IV, 101, Oldenburg to Pierre Carcavy, 2 January 1667/8 (trans from French by editors); see also Biagioli, 201.

It reminds us, too, that virtually any item of Boyle's writing exposes his enthusiastic use of literary genres (the familiar letter, the dialogue, the meditative essay, among others[20]) and rhetorical flourishes of all kinds, particularly those that frame investigative and theoretical discourse as social, courteous, and collaborative. The conventions that textualized bodily presence or mimicked orality in early-modern epistolarity may have been constructed to sustain the higher standing and register of the oral over the written; but I would suggest that what early-modern utopian correspondents also took from this convention was its dialogic quality—the letter as one half of an incompletely represented conversation or one voice in an imagined dialogue—as a way of sustaining their essentially collaborative ethos. The learned letters of the scientists often seem to fall between the Lipsian categories of *docta* (treatises) and *epistolae familiares* (familiar letters) (Dunn 1956, 154)—the content was learned and technical, but it had to be presented both familiarly and yet with ceremonious observances which recognized a specific interlocutor. The scientific letter was in effect one half, one speaking part, of a vocally imagined discussion, with some of the spontaneity of the spoken word and the implied mutuality of a dialogue's 'civil conversation'. The sense of this bifurcated generic category is well represented not just in correspondence but in a number of Boyle's other works, in which he addresses his nephew Richard Jones as Pyrophilus but does not allow him his own responding voice.[21] There is, in other words, a close connexion between the linguistic and social concerns of the discursive scientific work and the erudite scientific letter.

If civil observances and collaborative politeness are by some writers considered to be *de trop* in scientific discourse, they are by others championed as integral to science; and by the same token, plainness and eloquence more broadly are categories contested by scientific writers from Bacon onward. Boyle is fickle in the matter of rhetorical ornament and grace, whether social, civil, or intellectually performative – *ornari res ipsa negat, contenta doceri*, he says on the one hand ('the subject itself

[20] *A Proemial Essay touching … Experimental Essays* (1661), *The origin of forms and qualities* (1666–7), *Seraphick Love* (1659), *The Unsuccessfulness of Experiments* (1661), *The Usefulness of Natural Philosophy* (1663, 1671) are all addressed to Pyrophilus; *The Reconcileableness of Reason and Religion* (1675), *The History of Mineral Waters* (1685), *Accidents of an Ague* (1664), *Aerial Noctiluca* (1680) are addressed to other correspondents, sometimes in epistolary form; *A Discourse of Things above Reason* (1681) and *The Sceptical Chymist* (1661) are dialogues. He also wrote a fully fleshed-out fiction—*The Romance of Theodora and Didymus* (1687)—as well as a highly fictionalized memoir, *Philaretus in His Minority* (1648–9), and the fragmentary 'philosophical romance' *The Aspiring Naturalist* [n.d.].

[21] Demetrius discusses the letter as a half-dialogue or part-conversation. See Demetrius, 'On Style' in D.A. Russell and M. Winterbottom (eds), *Ancient Literary Criticism: the Principal Texts in New Translations* (Oxford: Clarendon Press, 1972), 211. See also Jean Robertson, *The Art of Letter Writing: An Essay on the Handbooks published in England during the Sixteenth and Seventeenth Centuries* (Liverpool and London: University Press of Liverpool/Hodder and Stoughton, 1942).

needs no ornament; it is enough that it be understood')[22]—and yet he recognizes that 'comparisons' (i.e., analogies) are 'in request' among natural philosophers. 'Some experience has taught me', he said, 'that such a way of proposing and elucidating things, is either more clear, or, upon account of its novelty, wont to be more acceptable, than any other to our modern virtuosi.'[23] In seconding Bacon's advocacy of 'lively representation' in the advancement of learning,[24] he admits that 'Proper comparisons do the Imagination almost as much Service as Microscopes do the Eye'.[25] But Bacon's lively representation of ideas by analogy and metaphor becomes, in much scientific writing, much more than striking comparative figures: it becomes a literally 'lively' introduction of fictional characters who speak of scientific abstraction. The conceit of *The Sceptical Chymist* (1661)—a chemical dialogue spoken by named characters who enact an elaborate narrative of intellectual courtesy in a garden on a beautiful summer day—is exactly the kind of imaginative service to science Boyle is thinking of: it presents a courteous gathering of disinterested seekers after truth, a personified version of the imagined or virtual scientific dialogue being conducted by letter from all over Europe by members of an 'invisible college'. Boyle prized the vocal conventions and shape of the dialogue: 'I sometimes took Pleasure, to imagine two or three of my Friends to be present with me at the Occasion, that set my thoughts on work, and to make them Discourse as I fancy'd Persons, of their Breeding and tempers, would talk to one another on such an Occasion.'[26] Learned correspondence offered him precisely these rhetorical, tropic, and especially dialogic features; it was, he said, a sort of *prosopopoeia* or impersonation, a 'conversation practis'd by Letters' and 'the next Contentment to the former by neerlyest approaching it'.[27]

Scientific or learned conversation among established savants has been described as fundamentally related to the prevailing local political structure—for example, Mario Biagioli and Jay Tribby have argued that the work and manner of discourse and exchange in the various Florentine academies were organized to advance and preserve an elite Tuscanness, to celebrate and consolidate Medicean power, and to function as a kind of adjunct to the ducal court; and that scientific manners particularly evolved in them in order to maintain good relations with princely patrons.

22 Boyle, *Proemial Essay* in *Certain Physiological Essays* (1661), 12. See also Susan M. Fitzmaurice, *The Familiar Letter in Early Modern English* (Amsterdam and Philadelphia: John Benjamins, 2002), 17–54.

23 Boyle, *Christian Virtuoso*, [A2v].

24 Francis Bacon, *The Advancement of Learning*, bk II, pt II, sect 5 *The Advancement of Learning* and *New Atlantis*, ed Arthur Johnston (Oxford: Clarendon Press, 1974), 73.

25 Boyle, *The Christian Virtuoso*, [A6r].

26 Boyle, *Occasional Reflections* (1665), B2r.

27 Boyle to John Mallet, 23 September 1653 in *The Correspondence of Robert Boyle*, eds Michael Hunter, Antonio Clericuzio, and Lawrence Principe, 7 vols (London: Pickering and Chatto, 2001) I, 146 (thereafter *BC*, followed by volume and page-number).

But I mildly disagree with that extension of the idea, which argues that this absolutist framework can, in a more corporate style, account for the civil behaviour of English scientists. Charles II was only a titular patron of the Royal Society. The Royal Society did not replicate monarchical or patronly authority, nor could it, since it was so carefully constructed, at least notionally, as a society of equals and had no single locus of authority other than the agreement of its fellows in discussion.[28] Moreover, a great many celebrated scientists and natural historians were working remotely and in isolation; and a surprising number of them, like Sir Thomas Browne, were not even members even if they were important correspondents. Scientific correspondence is—to adapt a remark of Adrian Johns's—rather a fictionally framed mode in the service of the fictional republic of letters and learning (Johns 1994, 13),[29] a nation specifically imagined as *external*, as self-consciously *outside* the prevailing political sphere, and existing in a notionally horizontal social landscape of mutual respect. But that republic was never more than a state of mind: it could not wholly absent itself from politics; and its conjured social topography was a fantasy of equality. It was a fiction, in other words, that begat fictions.

Oldenburg, who acted for many years as Boyle's agent with the London printers, was used to seeing Boyle's works through the press, and was thus more accustomed to receiving finished works. In 1666, Boyle apologized to Oldenburg for having sent him 'loos[e] papers' rather than fully worked-up scientific tracts. But he excused himself in the following rich metaphor:

> Yet the scruples I have are somewhat lessen'd, when I remember, that Men are serv'd and accommodated, not only by those Husbandmen, that once a year bring in whole wain-loads of Corn & Hogsheads of Wine, but also by Gardiners that do not wait for Autumn nor bring in at some such great & mature productions of their labour as Harvest and Vintage afford, but content themselves to be ever & anon furnishing the Markets wth Baskets of Roots & Herbs and Flowers and Grapes, and other fruit, and by the frequency & variety of these supplys make amends for the small Bulk of what they bring at a time.[30]

Boyle's tropes are always fascinating, and this one is peculiarly apt—scientific gleanings harvested from extended experimental and observational work are imagined as a rich bounty of food, wine, and seasonings, all as yet unassimilated into a dish or recipe, certainly still undigested. Latent in Boyle's culinary vision is the trope of parts or fragments still to be convened in a coherent order, an image of the

---

[28] See Biagioli, who makes the ingenious analogy between the Society/the *Philosophical Transactions*/Oldenburg, and princely power, with contributions to the *Philosophical Transactions* equivalent to gifts to a sovereign power (208–10); and Jay Tribby, 'Dante's Restaurant: The Cultural work of experiment in early modern Tuscany', in Ann Bermingham and John Beaver (eds), *The Consumption of Culture 1600–1800: Image, Object, Text* (London: Routledge, 1995), 319–37. David Kronick's assessment of what he calls 'corresponding societies' is a better model for the communications between various members of what would or had become the Royal Society (Kronick, 36–7).

[29] Johns reminds us that whatever was claimed for such collaborative and civil conventions, the ideals of openness, selflessness, and intellectual generosity did not always, or even usually, prevail (11).

[30] *OC* III, 145, Boyle to Oldenburg, May/June 1666.

fractured state of knowledge familiar from Bacon and his followers.[31] Oldenburg often described felicitously presented science as 'mak[ing] the mouth to water' (Avramov 1999, 193), and intellectual work in whatever form or state of completion as tasty sustenance was also a favourite conceit of John Evelyn's. He liked to imagine a 'deipnosophisse'—this is his term for a philosophical supper, adapted from Athenaeus's *Deipnosophistae* ('dinner-table savants')[32]—where men of learning exchanged their findings in person or by letter as a kind of breaking of bread. It is a conceit with undertones, of course, of the Platonic dialogue: casting the conceptual dialogue of epistolary transaction as a face-to-face symposium or 'deipnosophisse' naturally makes it especially congenial. Evelyn further elaborated the meeting of ideas as a rite of civil conversation, usually possible only under special conditions of privacy and tranquillity, when he proposed to Thomas Browne an ideal association of 'hortulan saints' or '*Paradisi Cultores*, persons of antient simplicity, … to be a society of learned and ingenious men'.[33] These sublime gardeners, he told Robert Boyle, would be

> Gentlemen who[,] … desir[ing] nothing more than to give a good example, preserve science, and cultivate themselves[,] … would resign themselves to live profitably and sweetely together … free from pedantrie and ill affectation … possessed of the most blessed life, that virtuous persons could wish or aspire to in this miserable and uncertain pilgrimage.'[34]

For Boyle's delectation these hortulan saints became desert fathers when Evelyn asked: 'is not this the same that many noble personages did at the confusion of the Empire … when St Hierome, Eustochion and others retir'd from the impertinencys of the world…?'[35] Evelyn made an elaborate plan of a college of scientists, for which he specified a dining schedule, and periods of 'conversation in the Refectory.'[36] Evelyn was still using this conceit almost a quarter century later when he described Dartmouth's expedition to Tangier in 1683 (which included Pepys) 'a Colledge, nay a[n] whole Universitie, all the Sciences, all the Arts, and all the Professors of 'em too … the ship that Athenaeus speakes of…'.[37]

But the Athenaean dream of a company of intellectuals in prolonged retirement together, in a 'recreative' or 'ludic' space (Nardo 2000, 130–1), free from 'impertinencies', was never realized except in a few unusual and mainly unplanned instances. Even the associations and clubs that sprang up all over Europe—various royal or ducal academies; private-house or regional gatherings; semi-institutional ones and

---

[31] See note 10 above.

[32] Evelyn to Pepys, 1 March 1686 and 7 January, 1695 in Bédoyère, 167, 256.

[33] Evelyn to Thomas Browne, 28 January 1659/60 in *The Works of Sir Thomas Browne* 4 vols, ed. Geoffrey Keynes, 2nd edn (Chicago: University of Chicago Press, 1964), 275.

[34] *BC* I, 366, Evelyn to Boyle, 3 September, 1659.

[35] *BC* I, 368, Evelyn to Boyle, 3 September, 1659.

[36] *BC* I, 366–8, Evelyn to Boyle, 3 September 1659.

[37] Bédoyère, 143.

other, vaguer assemblies; and even educational establishments[38]—were virtually all *urban*, mostly subject to political or religious interference; and when they could flourish in Evelyn's rural retirement it was only fitfully, during inclement political weather or the plague. And sometimes they did not flourish even then: Boyle, who could flee London whenever he liked, especially when pestilential visitors who sought him out as a sort of tourist attraction were interfering with his experimental life in town, only found that in his solitude in Dorset he was obliged to instruct the gardener and the ploughman in their duties, that projectors were digging up his pigeonhouse for saltpetre, and that a chemical furnace sent from London had been broken during shipment. Deepest Oxfordshire, where he retreated in 1665, was no better a place, he declared, 'perfectly disfurnished of all that is requisite for an experiment.'[39] The ideal conditions for experiment and discussion were, in short, hardly to be hoped for, but still he longed for a laboratory-retreat which would be 'a kind of *Elizium*' where he could at least practise experimental science in peace and in which he could be 'transported and bewitch'd … by the Delights I tast in it.'[40] Such impertinencies of the real world must have encouraged him to prefer the virtual community of letters, for it was usually only by letter that such an intellectual, experimental Elizium could be sustained.

The truth is that a physical gathering of scientists was, before the era of the university department and the governmental research lab, essentially impossible. Indeed, it may be that the first company of scientists ever to gather for an extended period in a remote place without many impertinencies was under Robert Oppenheimer at Los Alamos at the end of the second war, and even so this was supported only by a global crisis and heroic bankrolling by an emerging military superpower.[41] For the seventeenth-century natural philosophers, political, professional, familial, even microbial,

[38] Princely academies flourished in Rome, Florence, Paris, and London; the private-house gatherings included the Montmor Academy as well as those convened by Thevenot, de Thon, and Renaudot in Paris; the Collegium Curiosum in Altdorf (see note on Johann Christoph Sturm in *OC* III, 408), the Societas Christiana in Tubingen, the Academia Naturae Curiosum in Schweinfurt; in England the Great Tew Circle (of which the young Cowley had been a member), and the Towneley group in Lancashire were regionally based or at least managed; the early Gresham (or 'the Invisible') College (*OC* III, xxiv), and the Oxford group in the 1650s were never formally structured; even less formal were Gaspar Kalthoff's in Vauxhall (*BC* I, 178), the 'chemical council' which was said to meet near Charing Cross (*BC* I, 174), the herbalists and natural historians of Lime Street in London (described by Deborah Harkness, *The Jewel House: Elizabethan London and the Scientific Revolution* (New Haven: Yale University Press, 2007), 18, and even the special laboratory established for the poor in Durham (*BC* I, 292).

[39] *OC* II, 639, Boyle to Oldenburg, 9 December 1665.

[40] *BC* I, 83, Boyle to Lady Ranelagh, 31 August 1649.

[41] This atomic 'college' in the desert was in fact only maintained through Oppenheimer's assumption of the gatekeeper role: his scientists were mainly unmolested by the American military managers who were actually in charge because Oppenheimer fought all the administrative battles alone and on their behalf. See Kai Bird, *American Prometheus: the triumph and tragedy of J Robert Oppenheimer* (New York: Knopf, 2005); and Alice Kimball Smith, *Robert Oppenheimer: Letters and Recollections* (Cambridge, MA: Harvard University Press, 1980).

necessity stood in the way of such congregations for any meaningful length of time,[42] and Oldenburg found it difficult enough to keep even the chartered and constituted Royal Society going on a regular basis. 'Our meetings are very thin;' he reported to Boyle, 'and … our committees fall to ye ground, because it is not possible to bring people together.'[43] Some natural philosophers were perpetually and unwillingly rusticated, and could hope to sit down at the table of ideas only by letter. In 1646 Thomas Browne in Norwich apologized in the epistolary preface to *Pseudodoxia Epidemica* that although 'a worke of such concernment unto truth … did well deserve the conjunction of many heads and … some cooperating advancers … the privacie of our condition' had forced him to 'single and unsupported endeavours.'[44] Jeremiah Shakerley, a young mathematician and protégé of the Towneleys, wrote extensively from Lancashire to William Lilly in London. His letters are achingly ceremonious, possibly an index of his ignorance of the slightly easier epistolary style acceptable to English experimentalists,[45] with salutations which ought to have made Lilly blush: 'I may justly say of you as Kepler said of Tycho, Umbra fui sine te, te patre corpus ero [I was a mere shadow without you, through you, oh father, I will become substance] for in you the deadnesse of my hopes do liv, by you the powers of my minde do move, & from you the outward perspective of my selfe may [drawe] its being.'[46] He complained to Lilly that he lived 'in a County where good words are as rare as good wits: and ignorance seals the Lips of Complement … Happier had I beene, if I had never knowne any thing in the Mathematicall Sciences worthy knowledge[,] then knowing so much as I doe to bee debarred the use of it.'[47] John Beale, the eccentric ciderist whom Boyle and Hartlib referred to as '[our] Herefordshire philosopher', was in contact almost exclusively by letter and wrote voluminous screeds of fact and opinion in syntax and spelling erratic even by the standards of the day. His sedulous epistolary presence was a curse as well as a blessing: 'I wish,' wrote the weary Oldenburg to Boyle, 'that Dr Beale had digested his owne sense to you, and not commissioned me to cull it out of his letters here and there.' When experimental undertakings were so often either disturbed or disabled by distance and the absence of like minds, it is little wonder that the republic of learning wishfully held its philosophical suppers not at laden tables but in the virtual realm of correspondence.

The constantly-referred-to epistolary republic became the conceptual framework in which the scientists invested their dream of a nation dedicated to the advancement of learning because science's functional communal requirement found its next

[42] When the plague drove individuals out of London, they sometimes found themselves unexpectedly together at some country house or other with like-minded savants, as Robert Hooke, William Petty, and John Wilkins did in the plague summer of 1665.

[43] *OC* II, 235, Oldenburg to Boyle, 29 September 1664.

[44] *Pseudodoxia Epidemica*, 'To the Reader', 1.

[45] On this style, see Dunn, 152–4.

[46] Bodley MS Ashmole 423, fol. 117r.

[47] Bodley MS Ashmole 423, fols 114r–v.

best option in an epistolary tradition that sought to simulate presence textually; that this scientific community was a virtual rather than an actual nation deliberately blurs (or even plays up) the distinction between the literal and the metaphoric: 'Since you have now been *from* us for severall years,' wrote Oldenburg to John Winthrop in Connecticut in 1667, 'give us at last a visit by a Philosophicall Letter;'[48] and elsewhere he presents himself to correspondents as 'waiting upon [them] by letter'[49] as if epistolary representation were virtually the same as actual presence. Steven Shapin has noted 'the overwhelming importance of the letter as informal means of communication in contemporary philosophical activity'(Shapin 1987, 419). The informality of correspondence was no doubt striking in comparison with printed works, but I think it is essential to understand the letter as in other ways anything *but* informal. The natural philosophers had to fashion themselves, using the protocols, stylistic rota, and registers of epistolary propriety, as exemplary of the courtesy and conduct required for the prosecution of collaborative investigative projects.[50] For these reasons Boyle fretted about the salutations in his public letters in the *Philosophical Transactions*, that they might seem 'either too abrupt, or not so civil [...], as I desire my writings should be thought, as well as be.'[51]

What was the *literary* status of the early-modern letter? The somewhat *détendu* French astronomer Adrien Auzout thought that 'when one exchanges ideas by letter one does not look for eloquent and polished but plain and simple language...'.[52] James Howell praised the 'Familiar or Letter-missive' as peculiarly candid:

> *Speech* is the *Index*, *Letters Idea's* are
> Of the informing soul, they can declare,
> And shew the inward man as we behold
> A face reflecting in a Chrystal Mould.[53]

These are redactions of various well-rehearsed early-modern advices and manuals on the genre—Justus Lipsius advocating spoken rather than oratorical cadences in *Epistolica Institutio* (1587); Gassendi's influential letters (1630s) in the isogogic and instructional mode; Balzac's *Letters* (1624) in the style of the essay. Together with the theoretical writings of Demetrius, and the *epistolae* of Seneca, these models provided tonal and rhetorical templates for scientific letters. Instructions for early-modern epistolary style were, not surprisingly, often couched in courtesy or conduct manuals, an acknowledgment of the important relation between civil forms and the conduct of epistolary conversation. Such advice influences the specifically scientific versions of civility being promulgated and practised by Oldenburg, Boyle, and their correspondents, and had its own peculiar register within the spectrum of Renaissance

[48] Winthrop had in fact written numerous times, but the ships were lost. See Dorothy Stimson, 'Hartlib, Haak and Oldenburg: Intelligencers', *Isis* 31:2 (1940), 309–26, at 309.

[49] *OC* III, 525, Oldenburg to Winthrop, 13 October 1667.

[50] This point is also made by Shapin, 421–2.

[51] *OC* IV, 93, Boyle to Oldenburg, 29 December 1667.

[52] *OC* II, 518 Auzout to Oldenburg, 23 September 1665 (trans from French by eds).

[53] James Howell, *Epistolae Ho-Elianae* (1645), [A4v].

epistolarity, something between the more exposed, more familiar, and free-standing manner of the prose letters of Donne, who regarded letters as 'sacraments' of friendship and as essentially literary, as verse-epistles, and the more digested, moderate, Lipsian, less spontaneous quality of Joseph Hall (Corthell 1981, 416). 'Letters may[,] more than *History*[,] inclose the choicest learning,' said Howell. 'Choicest learning' couched in Howell's 'gentle and familiar' style is a formula for understanding early-modern scientific correspondence.[54]

Within an epistolary genre that partook both of the formal style of the treatise and the easier one of the familiar letter, the corresponding members could sit down at the table of learning, their epistolary presence almost, but not quite, material. And indeed, for all that Oldenburg and his correspondents complained about the state of the post and the constant delay or loss of letters in transit, it is clear that no material meeting of such a gathering would have been wholly desirable anyway, even were it possible. The maintenance of strict civility and cooperative generosity within his network could be better managed on paper by the adroit Oldenburg, a civility which in person might not have been successful among competitive and sometimes difficult personalities like Robert Hooke and Auzout. In other words, the trope of the ideal gathering is a rhetorical habit only partly based in real social practice; it was most powerfully felt via the letter and its cousin the dialogue, as a fictional or virtual *donnée*. All the impertinencies of the world and of personality argued for, rather than against, the value of virtual encounter, and the epistolary network is thus not a make-do arrangement, a pale mimesis of something it could never be; it is rather a rhetorical replication that solidified into its own reality, one that was as (or even more) significant than its original, an end in itself. Oldenburg's establishment of a house-style of conduct and expression for what we might call 'science-management' was thus heavily determined by presiding social and civil epistolary conventions; but the rhetorical consequences of these conventions developed, in the scientific letter, into a quite separate and specialized fiction of presence.[55]

How did the scientists imagine the rhetorical qualities of their correspondence? An amateur French chemist writing from Montpellier saluted Oldenburg almost mystically in 1666 as 'the chancellor of all the oracles pronounced in that celebrated academy in London', and one M. Bley, a member of the Montmor Academy, reportedly admired the Royal Society's 'sedate and friendly way of conference, as also ye gravity and majesticknes of [its] order.'[56] But Oldenburg and other fellows of the

---

[54] Howell, [A3v].

[55] Oldenburg's management of this development is well known: his primary function was to receive letters, digest them, and either respond to them, forward them, paraphrase them, or read them out to interested parties, and often to edit them for inclusion in the *Philosophical Transactions*. German by origin, his ability to conduct such business in Latin and at least five modern languages [English, German, French, Dutch, Italian] was crucial to the success of the fledgling society and of interactive European science. The profound significance of Oldenburg's labours and his peculiar talents and qualifications to perform them is also well known.

[56] *OC* II, 235, Oldenburg to Boyle, 22 September, 1664.

Royal Society often had to coach various correspondents vigorously in the rules of scientific discussion, and especially of epistolary exchange, and to act as disinterested mediators in disputes. When Eckhard Leichner of Erfurt wrote expressing rather scholastic notions of scholarly dispute, Oldenburg explained that the Royal Society was not concerned with scholastic or theological learning, but rather with 'knowledge of nature … by means of observation and experiment.' Leichner did not respond.[57] When Auzout and Johannes Hevelius of Danzig publicly disagreed in their observations of a comet which appeared in 1664–1665, Oldenburg refereed the debate by putting the evidence of each before the Society. It found in favour of Auzout, a decision fraught with social complications for the Society in publicly disputing Hevelius's credit. The tenacious and somewhat excitable Hevelius was naturally reluctant to accept this verdict, and did so only when Oldenburg seemed to suggest the possible but improbable solution of there having been two 'Phaenomena' in the sky.[58] Here we see Oldenburg acting as impartial conduit of information and as adjudicator, but also as ambassador, of the Royal Society and of the more abstract general congregation of 'learned opinion'.[59] Oldenburg constantly solicited his correspondents for news of recent investigations, and was solicited in return. It could be a thankless undertaking: Hevelius overrode the Royal Society's arbitration of his case and kept trying to re-open the dispute despite Oldenburg's gentle attempts to stop him.

Hevelius continued to defend his own claim privately, while publicly submitting to the Royal Society's judgment, apparently not quite *au fait* with the canons of civil exchange and collaborative enterprise. For example, Oldenburg had to explain to him why Fellows were enjoined to proclaim their membership of the Royal Society in their printed works, that strength lay in corporate, collaborative achievement, in a certain philosophical selflessness which Hevelius clearly had some difficulty comprehending.[60] All natural philosophers were equal 'citizens of our scientific community', he reminded him.[61] And Hevelius was not the only one in need of instruction. To the triumphant Auzout Oldenburg also wrote during one of his *contretemps* with Hooke: 'Surely, sir, it is indeed the right way to manage a correspondence between two worthy men and fine minds, when each expresses to the other his thoughts and discoveries in a frank and polite way, without offence given or taken…'[62] Auzout in any case habitually infringed on the laws of correspondence by his sheer, self-confessed

[57] *OC* II, 111, Oldenburg to Leichner, 23 September 1663.

[58] *OC* III, 313, John Wallis to Oldenburg, 19 January 1666/7.

[59] Biagioli regards Oldenburg as inhabiting the 'princely' role within the Royal Society, equivalent to the royal patron in Italian academies. But this attempt to analogize the Royal Society and its Tuscan counterparts neglects Oldenburg's tone and self-description as servant, conduit, and spokesman rather than as autocrat or power-broker (Biagioli, 210).

[60] *OC* II, 396, Hevelius to Oldenburg, 22 May 1665 (trans from Latin by eds).

[61] *OC* III, 76, Oldenburg to Hevelius, 30 March 1666 (trans from Latin by eds).

[62] *OC* II, 441 Oldenburg to Auzout, 23 July 1665 (trans from French by eds).

laziness: 'I am seldom in the mood for writing letters', he explained without apology,[63] although he knew he had thereby forfeited the right to expect punctual replies, or any reply at all, from anyone else.[64] And when Auzout published a private letter written to him by Oldenburg about recent work by Robert Hooke, Oldenburg was naturally displeased, not least because it landed him in hot water with the prickly Hooke. But Auzout was unrepentant: 'I see little difference between printing scientific matters contained in letters, and showing these same letters to those [persons] learned in these matters … one writes, well aware of the fact that it will be shown to many learned men.' 'There is a great difference,' he went on, 'between an eloquent discourse and some[thing] which has been written … for oneself alone which one would never print without the author's permission …'[65] And, despite this failure of civility, both in publishing Oldenburg's letter and then lecturing him about the propriety of having done so, Auzout had a point: in the virtual world, the learned letter stands in for the social presence of its writer; it is no more private than an intervention in a meeting of a learned society. The emerging concept of privacy is one which was still unstable at this point, and Auzout's casual refusal to acknowledge it is perhaps key to understanding the regulation of learned letters.[66]

The academies and clubs were the virtual, epistolary republic reified. But perhaps because they were not virtual, they conducted their civil conversation within more overtly political conventions: despite their lofty utopianism, they often behaved like states or national entities. They regarded their natural civil relations to be with other such groups, or with reigning monarchs and petty despots, and they conducted their communications in formal patterns reminiscent of international diplomacy. There were comings and goings between these groups, travelling scientists urged by their friends at home to effect introductions to the local society of savants, to establish friendly relations by the exchange of courtesies and of new scientific information, specimens, and books. The Royal Society planned to receive Charles II with a 'philosophicall entertainment'[67]; and Oldenburg urged Boyle, as the Society's leading aristocratic natural philosopher, to establish and maintain contact with Prince Leopoldo and the Accademia del Cimento on behalf of the Society. The usual precise and formal civilities were always emphatically asserted and strictly maintained. The often embattled circumstances of seventeenth-century *realpolitik* directly opposed the utopian model of scientific cooperation, and prompted such exact observances within learned circles. Perpetual war and foreign threats shadowed the horizon, but the scientists could imagine (or delude themselves with) an insulating

[63] *OC* IV, 65, Auzout to Oldenburg, 19 December 1667 (trans from French by eds).

[64] *OC* IV, 227, Auzout to Oldenburg, 7 March 1667/8 (trans from French by eds).

[65] Auzout to Oldenburg, 23 September 1665 (*OC* II, 518) (trans from French by eds).

[66] On early-modern privacy, see Ronald Huebert, 'Privacy: The Early Social History of a Word', *Sewanee Review* 105 (1997), 21–38. On privacy and early-modern correspondence, see Schneider, 24 and throughout.

[67] Oldenburg to Boyle, 2 July 1663 (*OC* II, 78).

ideology of ordered intellectual endeavour which rigorously excluded religion, politics, and the ambitions of empire. Even when they had, like the Royal Society, evolved out of nationalist ambitions, the academies, and certainly individual scientists and philosophers who conversed privately with one another, resisted national alliances and battles by maintaining contact outside prevailing political conditions. Spinoza, in principle an enemy alien, wrote to Oldenburg in 1665 to say that the wars had even moved him to 'let everyone live according to his own ideas, and let those who *will*[,] die for their own good, so long as *I* am allowed to live for the truth.'[68] 'O for a Paradise in which wee might retreate from the noyse of Trumpet and drum,' John Beale wrote to Hartlib in 1659.[69] In fact, Beale had already found that retreat in the learned republic of letters.

In 1663, when Samuel de Sorbiere, the French physician and natural philosopher, was thought to have represented himself without permission as the designated ambassador to the Royal Society from the Montmor Academy, his colleagues were outraged. Another member, Pierre Petit (himself a government official under Richelieu), wrote to Oldenburg complaining of this shocking breach of etiquette. Oldenburg tried to assure Petit that Sorbiere had *not* in fact claimed to be visiting London except as a private individual, but to no avail; even though Sorbiere declared himself 'carried away by admiration for the Royal Society of London Philosophers,'[70] the suspected injury to manners lost him many friends both in Paris and London. What seems a trifling breach of civility shows the learned academies acting as nations or city-states with accredited diplomats and the formal relations of potentially hostile parties—parties that, after all, were not immune from territorial disputes over priority of discovery or invention—and yet their relations, at least in the correspondence, were studiedly careless of international ones. 'This company,' Oldenburg explained to Auzout, 'holds the opinion that political and national difference ought not to impede philosophical exchange nor shut the door upon the appreciation of the sciences and virtue.'[71] And Sorbiere, in defending himself, claimed to be (in Oldenburg's paraphrase) 'awkward and ignorant of the sublimer sciences, of the measure of men, and of politics especially.'[72] But it was a state of mind, promulgated through letters, that was nevertheless vulnerable to real-world interventions: the ire of Thomas Sprat, among others in London, was such that Sorbiere's supposed infringement in the virtual world brought consequences in the real one when he was arrested and banished from Paris by the French king to forestall a diplomatic incident. (It should not be forgotten that Sorbiere also published an account of his

[68] *OC* II, 541, Spinoza to Oldenburg, September/October 1665 (trans from Latin by eds); my emphases.

[69] Beale to Hartlib, 4 November 1659, quoted in Charles Webster, *Utopian Planning and the Puritan revolution: Gabriel Plattes, Samuel Hartlib, and Macaria* (Oxford: Wellcome Unit for the History of Medicine, 1979), 7.

[70] See the extended correspondence on this matter in *OC* II, especially 134 and 267.

[71] *OC* III, 141, Oldenburg to Auzout 24 May 1666.

[72] *OC* II, 267, Oldenburg to Boyle (quoting Sorbiere), 20 October 1664.

visit to England in which he rubbished English food and English hospitality, a civil breach which Sprat found unforgiveable.[73]) On another occasion, when Pierre Carcavy, the French royal librarian and a noted mathematician, suggested that he and Oldenburg enter into correspondence as representatives of their respective scientific academies, the Royal Society judged the letter 'not to be written in the name of that [French] academy to the society, but only by a single member thereof, expressing his desire to correspond; Mr Oldenburg should only as from *himself* thank him for his offer, and entertain a correspondence with him upon philosophical matters.'[74] Oldenburg, the very architect of scientific courtesy, complained privately to Boyle that he was dissatisfied with this fastidious decision: 'I see, Punctilio's retard and obstruct much good, both publike and private. But I must submit.'[75] Oldenburg had naturally been keen to make strong links with the French Academie, and an official correspondence with a mathematician with valuable royal connexions would have allowed him to forge them. Oldenburg was clearly impatient of the counterproductive institutional dignity and *amour-propre* of the learned societies, which he himself had in some measure fostered.

Institutional dignity contributed to another rhetorical problem, that of representation: how, in epistolary transaction, were scientists to manage the rhetorical difference between actual and epistolary presence? As we have seen, the receipt of a letter was clearly analogized as a polite encounter (as Oldenburg had regarded the prospect of a letter from Winthrop as a 'visit'), and correspondents could, via their missives, have a presence that could be almost theatrical. In the 1690s James Petiver, a well-connected London apothecary, was publishing in serial pamphlets what he called the *Musei Petiveriani*, a description of his renowned *hortus siccus*, or collection of dried plants. In several of these he makes acknowledgement of the many correspondents who contributed specimens and information to the collection by naming them all in a sort of Cowleyan prosopographical list from *Plantarum*, almost like a list of *dramatis personae*, and one that gives a strong sense of the group enterprise among people who, though rarely or never in each other's company, and indeed in this instance far-flung all over the world, are 'present' within the enterprise of collection and publication.[76]

But injury and insult, too, could be as keenly taken in absence. Boyle was told of a certain unnamed physician who had sourly disputed Hevelius's findings on blood transfusion during a meeting of the Royal Society. Oldenburg reported:

> I cd not but take him afterwards aside, and represent to him, How he would resent it, if he should communicate upon his owne knowledge an unusual Experiment to ye Curious at Dantzick, and they in publick brand it wth ye mark of falsehood: that such Expressing in so publick a place, and in so mixt an assembly, would certainly prove very destructive to all philosophical commerce.[77]

---

[73] Thomas Sprat, *Observations on Monsieur de Sorbier's Voyage into England* (1665).

[74] *OC* IV, 158–9 (editors' note).

[75] *OC* IV, 207, Oldenburg to Boyle, 25 February 1667/8.

[76] James Petiver, *Musei Petiveriani* [3rd pamphlet] (1699), fol. 101 and following (in BL MS Sloane 3330).

[77] *OC* IV, 27, Oldenburg to Boyle, 10 December 1667.

There were home-grown examples of such failures of courtesy, as Oldenburg had reason to know only too well: Robert Hooke had an ill-tempered dispute with him about his invention of the spring-powered watch, to which Hooke seemed to have been pipped by Christiaan Huygens; but as Oldenburg reported to Boyle, Hooke 'hath too slender thoughts of all what comes from abroad of a philosophicall nature, or is done by strangers.'[78] The intractable personalities of certain correspondents were a very good reason for preserving and fostering virtual meetings instead of face-to-face encounters.

Sometimes letters could resolve themselves out of the virtual world quite unexpectedly, and cease to be merely textual markers of what might otherwise have been oral discussion. As well as acting as Boyle's translator and agent, Oldenburg was passing to Joseph Williamson, working for Lord Arlington's intelligence office, anything that came to him in his voluminous foreign correspondence that might be useful to the English government (especially anything concerning the French and the Dutch) in exchange for postal discounts. Henri Justel in Paris and Baruch Spinoza in Holland were among his many sources of hearsay and opinion on the ground, and it was because of this network of information that Oldenburg was subjected to the most damaging incivility of all, imprisonment in the Tower in 1667, apparently for lamenting by letter to Justel the atrociously ill-prepared state of the English navy, which had recently been surprised by the Dutch in the Medway. For this remark (which was mere fact, as everyone knew), Oldenburg was detained two months as a suspect foreigner conversing with foreigners. The scientific ideal of liberal conversation across national and political boundaries was *not* in fact robust enough to withstand such an intervention; and it was the intercepted correspondence itself—the *material* evidence of the virtual world—which implicated him. It is significant, too, that despite their fine exclusion of politics from their scientific circle, not one of his Royal Society colleagues visited him, except the redoubtable John Evelyn whose senatorial attitude dismissed any danger to himself.[79] Oldenburg was eventually released without charge, but the Tower incident, which arose out of the 'innocent' epistolary transactions of his virtual learned republic, was a rough reminder that this intellectual nation had undefended borders, and the ideal of internal emigration to a country of learning could only barely be sustained within the political realities of Restoration England.

John Evelyn's idea of heaven was, he explained to Pepys in 1701, one 'where those whose refined and exalted nature makes capable of the sublimest mysterys, and aspire after experimental knowledge … shall meete with no prohibition of what is desireable, no serpent to deceive, none to be deceived. This is, Sir, the state of that Royal Society above, and of those who shall be the worthy members of it.'[80] The conclusions of old age were perhaps more satisfactory to Evelyn than to most;

[78] *OC* II, 248, Oldenburg to Boyle, 6 October 1664.

[79] Guy de la Bédoyère, *Particular Friends*, introduction, 14.

[80] Evelyn to Pepys, 10 December 1701 in Bédoyère, 286.

and utopian schemes are as a rule notoriously vague on the nuts and bolts of subsistence. The exceedingly utopian Royal Society, in accepting Oldenburg's heroic efforts on its behalf, had been perilously slow to provide him a desperately needed salary to support his work, as if life itself were 'virtual'. For this reason he could not but enter into other lucrative arrangements, and it was one of these that led to his unfortunate arrest. He complained that his obligingness as general secretary was too readily importuned, that the Royal Society below and its commonwealth of learning were sustained only *in sudore vultus*. For Oldenburg (no scientist) his distinctive gifts in the rhetorical, interpretive, managerial, and commercial—all key, pragmatic skills of the real world which his scientific colleagues often lacked—made an interesting and fortunate combination in which worldly and utopian were joined by his incessant labour, and through which he manipulated the fragile rhetorical framework on which rested the possibility (if not always the reality) of collaborative science and civil intellectual exchange in an uncomfortable world; through correspondence and its functional rhetorical patterns, he could regulate the learned economy of science and spin the fable of that Royal Society above.

References

Avramov, Iordan. 1999. An apprenticeship in scientific communication: The early correspondence of Henry Oldenburg (1656–63). *Notes and Records of the Royal Society of London* 53(2): 187–201.

Biagioli, Mario. 1996. Etiquette, interdependence, and sociability in seventeenth-century science. *Critical Enquiry* 22: 193–238.

Corthell, Ronald J. 1981. "Friendships Sacraments": John Donne's familiar letters. *Studies in Philology* 78(4): 409–25.

Dunn, E.Catherine. 1956. Lipsius and the art of letter-writing. *Studies in the Renaissance* 3: 145–56.

Johns, Adrian. 1994. The idea of scientific collaboration: the "man of science" and the diffusion of knowledge. In *Commercium litterarium: Forms of communication in the republic of letters, 1600–1750*, ed. Hans Bot and Françoise Waquet, 3–22. Amsterdam/Maarson: APA-Holland University Press.

Kronick, David. 2001. The commerce of letters: networks and "invisible colleges" in seventeenth- and eighteenth-century Europe. *The Library Quarterly* 71(1): 28–43.

Nardo, Anna. 2000. A space for academic recreation: Milton's proposal. In *The reason of church government* in *Literary circles and cultural communities in Renaissance England*, ed. Claude J. Summers and Ted-Larry Pebworth, 128–47. Columbia/London: University of Missouri Press.

Schneider, Gary. 2005. *The culture of epistolarity: Vernacular letters and letter writing in early modern England, 1500–1700*, 109–42. Newark: University of Delaware Press.

Shapin, Steven. 1987. O Henry. *Isis* 78: 417–24.

Chapter 10
Debating the Faith: Damaris Masham (1658–1708) and Religious Controversy

Sarah Hutton

10.1 Debating the Faith: Damaris Masham (1658–1708) and Religious Controversy

Damaris Masham was one of a tiny handful of seventeenth-century women who published philosophical writings, and like many women of this period, she also engaged in correspondence.[1] Her two books, both small octavo volumes, were published anonymously: *A Discourse Concerning the Love of God* (1696) and *Occasional Thoughts in Reference to a Vertuous or Christian Life* (1705). Both her books (as their titles indicate) and her letters, contain discussions not just of philosophy but of diverse religious matters, including such topics as the relationship of reason to faith, religious toleration and controversies surrounding the doctrine of the Trinity. These were all topical issues in the last decades of the seventeenth century. In this essay I examine Damaris's interest in religion, as expressed in her letters, to see what they reveal about her religious views, her knowledge of theology and the extent to which she engaged in debate on matters of faith. Her correspondence with John Locke is a major source for this, but I am not concerned here with examining the extent to which her religious views coincide with his, or whether she was influenced by him (or, indeed, the reverse). Rather, as I hope to show, the matters which he debated with friends and foes were also matters on which she had her own views.

[1] For the philosophy of Damaris Masham, see Sarah Hutton, 'Damaris Cudworth, Lady Masham: between Platonism and Enlightenment', *British Journal for the History of Philosophy* 1 (1993): 29–54, James G. Buickerood, 'Masham, Damaris', in Andrew Pyle (ed.), *The Dictionary of Seventeenth-Century Philosophy*, 2 vols (Bristol: Thoemmes Press, 2000), 2: 559–62, Jacqueline Broad, *Women Philosophers of the Seventeenth Century* (Cambridge: Cambridge University Press, 2003). Details of her correspondence are given below.

S. Hutton (✉)
Adran Saesneg, Aberystwyth University, Penglais, SH23 4QY Aberystwyth, UK
e-mail: sfh@aber.ac.uk

A. Dunan-Page and C. Prunier (eds.), *Debating the Faith: Religion and Letter Writing in Great Britain, 1550-1800*, International Archives of the History of Ideas 209,
DOI 10.1007/978-94-007-5216-0_10,

Also important for understanding the extent and character of her participation in debates about religion are the handful of letters which she exchanged with Locke's friends, Philip Van Limborch and Jean Le Clerc. These not only shed light on her knowledge of controversial issues, but they give us some sense of her involvement in debates about religion during the time of her closest contact with Locke. Her correspondence with Le Clerc and Van Limborch add an international dimension to her epistolary commerce, marking her entry to the Republic of Letters. (In this respect, her correspondence is quite unlike that of any other English women letter-writers of this period).[2] This, in turn, drew her into debate with two of the brightest minds of European Protestantism, Gottfried Wilhelm Leibniz—with whom she corresponded directly—and Pierre Bayle, with whom she crossed swords (by proxy through Jean Le Clerc). For the purposes of this paper, my main focus will be her letters to Locke, Le Clerc and Van Limborch. There is more to say about her correspondence with Bayle and Leibniz, both of which are the subject of other studies (Colie 1957; Phemister 2004; Sleigh 2005; Phemister and Smith 2007).

Since most women philosophers of the seventeenth century either had no opportunity to publish anything, or published very little, letter-writing was an important means for women to engage in philosophy, and a number of them did so by corresponding with male philosophers. Notable examples of such philosophical epistolary exchanges are Anne Conway's correspondence with Henry More, Mary Astell's with John Norris, and the correspondence of the woman who forms the subject of my paper, Damaris Masham, who corresponded with two of the foremost philosophers of her time, John Locke and Gottfried Wilhelm Leibniz.[3] It is no accident that religion should be prominent in the letters of these women. This is partly for the reason that religion figures in most female correspondences. In particular, religion is very prominent in some of the best known correspondences between men and women in the period, several of which take the form of spiritual counselling or have a strongly spiritual dimension to them. Examples are the correspondence of Simon Patrick and Elizabeth Gauden, the 'seraphic' friendship between John Evelyn and Mary Godolphin (Harris 2003).[4] The spiritual content of correspondences like these suggests that religion was an acceptable basis for men and women to engage in epistolary relationships. The correspondences of several female philosophers contain an element of spiritual counselling. The Norris-Astell exchange, published as *Letters concerning the Love of God* with its strong religious focus is of this variety.

---

[2] One exception is Dorothy Moore, who married John Durie. See *The Letters of Dorothy Moore 1612–1664*, ed. Lynette Hunter (Aldershot: Ashgate, 2004).

[3] See *The Conway Letters: the Correspondence of Anne, Viscountess Conway, Henry More and their Friends, 1642–1684,* ed. Marjorie Nicolson, rvd Sarah Hutton (Oxford: Clarendon Press, 1992); Norris, *Letters Concerning the Love of God*, ed. E. Derek Taylor and Melvyn New (Aldershot: Ashgate, 2005); G.W. Leibniz, *Die Philosophischen Schriften von Gottfried Wilhelm Leibniz*, ed. C.I. Gerhardt, 7 vols (Berlin, 1875–1890), vol. 3.

[4] For other examples of intense, but chaste relationships, known by the term 'Seraphic Love', see Cornelia Wilde, 'Seraphics, Ideals and practices of Neo-Platonick Love and Friendship in Seventeenth-Century England', unpublished PhD thesis, Humboldt University (2009).

Even in Anne Conway's correspondence with Henry More we never lose sight of the fact that he was a man of the cloth. Damaris's correspondence is certainly not in this mould. She corresponded not with spiritual counsellors, but with active members of the Republic of Letters. But there is a 'Seraphic' aspect to her relationship with Locke (Goldie 2004).

Another reason why religion figures prominently in the writings of female philosophers is that, in this period, the dividing line between religion and philosophy was not clear cut, it is not unusual to find religion and philosophy intertwined. After all many aspects of religion are cognate with philosophical issues, or lend themselves to philosophical discussion—particularly moral questions and metaphysical topics, such as the existence of God. But the interest in religious matters taken by women philosophers can also be explained by the fact that where religious topics are open to philosophical analysis, they offered a way of opening up philosophical questions. Thus discussion of religion was a route to philosophising. And this was particularly important for those who, like women, did not have access to the discursive spaces open to educated men. As Catharine Macaulay claimed in the next century, 'All true religion may, undoubtedly, be styled philosophy', or as Damaris Masham put it, 'Religion is the Concernment of All Mankind; Philosophy as distinguish'd from It, onely of Those that have a freedome from the Affaires of the World'.[5] Discussion of religion on this definition is ancillary to philosophy, serving as a substitute for philosophy for those with less freedom including women.

A major limitation about relying on letters as sources, is that only rarely do complete correspondences survive. The extant Conway-More correspondence, for instance, consists largely of his letters to her. With Damaris Masham, the obverse is true of her correspondence with Locke: it is mainly her letters to him which survive—thanks to his meticulousness about keeping and filing the letters he received. There is the further difficulty that there are no letters for the last decade of his life, after he took up residence at her house, apart from one or two written on his occasional visits to London. Since the period of Locke's residency at Oates was also the period of sharp religious controversy in England, into which Locke was reluctantly drawn, the question arises of how far Damaris Masham was party to these debates. Frustratingly, however, this is also the period when the correspondence between Lady Masham and John Locke all but ceases. To get a sense of the conversations to which she may have been party, we have to rely on the correspondence of others. Locke's letters to and from other people during this time, especially his correspondence with Le Clerc and Van Limborch in the Netherlands, frequently mention Lady Masham. This raises the intriguing possibility, which I shall address below, that she was a 'reader over the shoulder' of Locke's correspondence, and possibly even a participant in the debates. Furthermore, her own letters to these two friends of Locke give some insight into her own beliefs, and her knowledge of theology,

[5] Catharine Macaulay, *A Treatise on the Immutability of Moral Truth* (London, 1783), 290. Damaris Masham, letter to Locke, *The Correspondence of John Locke*, 8 vols, ed. E.S. de Beer (Oxford: Clarendon Press, 1976–1989), 3: 432, letter 1040.

since, in both cases, Lady Masham takes occasion to underline theological common ground between herself and her correspondent. These letters also shed some light on how Lady Masham intervened in religious debates, through the way in which she managed her interventions in the religious and philosophical debates of the Republic of Letters.

10.2 Lady Masham's Correspondents

Damaris Masham's correspondence with John Locke is by far the largest and longest of her extant correspondence. It comprises some 45 letters, mainly from her, written between 1682 and 1697.[6] Although she does deal with faith-related philosophical issues (such as proof of the existence of God and the role of reason in matters of religion) her references to religion in her letters to Locke are more commonly light-hearted (Hutton forthcoming). Her letters also give us a sense of how her letter-writing fitted into her daily life, giving a vivid impression of just how hard it was for her to balance her intellectual interests with her role as lady of the manor, with all its domestic and social demands on her time (Hutton 1993). Locke is a key figure in both Lady Masham's philosophical life, and also in her personal life (Simonutti 1987; Hutton 1993; Goldie 2004).[7] The question of how far her philosophy is indebted to his is a matter for debate, and need not concern us now. But there is no question that he played a significant role in enabling her to pursue her philosophical interests by providing a stimulating intellectual environment—either through letters or, by personal contact. As we shall see, they certainly also had much in common from a religious point of view, though we should not assume that they agreed on everything. Her friendship with Locke brought Lady Masham into contact with wider horizons, especially intellectual horizons, beyond High Laver in rural Essex where she lived as a married woman. When Locke came to live at her home there, it became not just the focal point for his English circle, but also an outpost of the Republic of Letters (Simonutti 2008) including, of course, friends and acquaintances from his period of exile in the Netherlands.

The first of these was Philip Van Limborch (1633–1712), the Dutch Remonstrant theologian and sometime professor of Theology at Amsterdam.[8] Van Limborch and

[6] For Damaris Masham's correspondence with Locke, see Locke, *Correspondence*, vols 2–6.

[7] The significance of Locke for Lady Masham's intellectual life does not mean that she did not have other resources through which to pursue her interests. As Margaret Ezell pointed out, female networks were important too; Margaret Ezell, '"Household Affaires are the Opium of the Soul": Damaris Masham and the Necessity of Women's Poetry', in Barbara Smith and Ursula Appelt (eds), *Write or Be Written: Early Modern Women Poets and Cultural Constraints* (Aldershot: Ashgate, 2001), 49–65. So were individuals: she was a friend of Lady Ranelagh, sister of Robert Boyle (Locke, *Correspondence*, 3: 278, 294, letters 967 and 975).

[8] For the Van Limborch letters see Amsterdam University Library, MS M31c. One is printed in Abraham des Amorie van der Hoeven, *De Joanne Clerico et Philippo a Leenborch Dissertationes duæ. Adhibitis epistolis aliisque scriptis ineditis, scripsit, atque eruditorum virorum epistolis nunc*

Locke had become acquainted while he was in exile in the Netherlands, where they had extensive discussions on religious matters, which they continued by letter after Locke's return to England. Lady Masham's extant correspondence with Van Limborch dates from 1704 to 1705, after Locke's death.[9] Van Limborch's letters are written in Latin, hers in French (translated by Pierre Coste). Lady Masham was prompted to write to Van Limborch to gather information for the biographical sketch of Locke on which she was engaged, wishing to find out more about Locke's residence in the Netherlands during his exile. Lady Masham's connection with Van Limborch was cemented at a personal level, and developed in other ways before Locke's death. Van Limborch had been a friend of Damaris Masham's father, Ralph Cudworth, many years previously, and they had exchanged letters which show that, as anti-Calvinists, they had much in common theologically.[10] Long before they were in direct contact by letter, they were delighted to rediscover the Cudworth connection. When Van Limborch's son, Francis, visited Locke, Lady Masham's warm hospitality left an abiding impression on the young man who remained in contact.[11]

The second member of Locke's Netherlands circle with whom Lady Masham corresponded was the Swiss scholar, theologian and editor Jean Le Clerc (1657–1736). There are three extant letters from her to Le Clerc.[12] Another letter is mentioned by both Pierre Coste and Le Clerc, which Le Clerc forwarded to Bayle.[13] Le Clerc was a vital link to the intellectual circles both in The Netherlands, where he lived, and in Europe, through the journals which he edited (*Bibliothèque universelle* and *Bibliothèque choisie*). Le Clerc did much to disseminate the philosophy of both Locke and Cudworth to an European audience especially through his *Bibliothèque choisie*.[14] The original reason for their contact concerned Ralph Cudworth, a copy of whose *True Intellectual System of the Universe* Lady Masham sent him in 1699.[15] Le Clerc subsequently published extracts from this in translation in the *Bibliothèque choisie*, sending Lady Masham copies of the relevant volumes.[16]

primum editis auxit (Amsterdam, 1843). See also Luisa Simonutti 'Religion, Philosophy, and Science: John Locke and Limborch's circle in Amsterdam' in James E. Force and David S. Katz (eds), *Everything connects: In Conference with Richard H. Popkin* (Leiden: Brill, 1999), 295–324.

[9] Amsterdam, UBA MS III D.

[10] Cudworth to Limborch, Amsterdam, Universiteits Bibliotheek, MS M.21.c. Cudworth sent Van Limborch a copy of *The True Intellectual System.*

[11] For Francis Van Limborch's letters see Locke, *Correspondence*, vols 6–8.

[12] There are four, if her account of Locke's life is included. These are printed in Jean Le Clerc, *Epistolario*, ed. Mario Sina, 4 vols (Firenze : Olschki, 1987–1997). Letters 342, 364, 380, 2: 389–91, 445–7, 497–517.

[13] Coste to Le Clerc, *Epistolario*, 4: 527–8, letter 383; *Bibliothèque choisie*, 7: 242–7.

[14] *Bibliothèque choisie* 28 vols (Amsterdam, 1703–1713). This is a continuation of *Bibliothèque universelle et historique* 25 vols (Amsterdam, 1686–1693).

[15] Le Clerc sent his thanks via Locke; Locke, *Correspondence* 6: 541, letter 2531.

[16] *Bibliothèque choisie*, 2, items 1&2 (1703); vol. 4, item 1 (1705); vol. 5, item 2 (1705) ; vol. 7, item 1 (1705); vol. 8, item 1 (1706). The letters which accompanied Le Clerc's gifts which are mentioned in letters to Locke (*Correspondence*, 6: 211 and 323, letters 3468, 3559) do not survive.

The first extant letter to Le Clerc from Lady Masham thanks him for publishing the first extract and for sending her a copy of the volume in which it appeared (her son, Francis Cudworth Masham also wrote to thank him—his letter, written in French, accompanied hers). Lady Masham's contact with Le Clerc antedates Locke's death, but continued after it for partly the same reason as her continuing contact with Van Limborch, *Éloge* of Locke which she researched and wrote, and which Le Clerc published in the *Bibliothèque choisie* in 1705.[17] In the same year Le Clerc published a review of Coste's translation of her *Discourse on the love of God*.[18]

Of course, the interest which any of these correspondents had in Lady Masham derived chiefly from her privileged contact with Locke. This is also true of her third foreign correspondent, Gottfried Wilhelm Leibniz, although he never managed to get into correspondence directly with Locke. The Leibniz-Lady Masham correspondence, which dates from 1703 to 1704 is a set of 12 letters from which both sides of the correspondence survive. These letters are mainly philosophical, focusing on Leibniz's theory of pre-established harmony. But the discussion has a religious dimension, especially as concerns God's relationship to the world.

Another foreign contact who should be mentioned, is Pierre Coste, who was engaged as tutor to Lady Masham's son in 1697. It was probably through Locke that he acquired this position. There are no extant letters between Lady Masham and Pierre Coste—if they were in correspondence, their letters do not survive. However, through his role as translator, Coste was pivotal in the European dissemination not just of Locke's philosophy, but also of Lady Masham's, for it was he who translated her *Discourse on the Love of God*, which was published in Amsterdam in 1705 and reviewed in *Bibliothèque choisie*.[19]

Lord Shaftesbury was another Lockeian connection of Lady Masham's but the handful of her extant letters to Lord Shaftesbury and to Samuel Clark have neither philosophical nor religious interest, as they concern items of mundane business.

Le Clerc was very pleased with the response to his publication of Cudworth in *Biblotheque choisie*. 'Cudworth, he told Locke had been 'd'un grand secours' for him, 'Tout le monde en a lu ici les Extraits avec plaisir; excepté quelques entêtez *cartesiens* ou Spinosistes, qui son ennemis de tout ce qui est different de leurs systemes' (Locke, *Correspondence*, 8: 35, letter 3319).

[17] 'Eloge de feu Mr Locke', *Bibliothèque choisie*, vol. 6, item 11 (1705).

[18] *Traité de l'amour divin* (Amsterdam, 1705); *Bibliothèque choisie* (1705), vol. 4, item 10, 383–90 Republished in 1715 as *Discours sur l'amour divin, où l'on explique ce que c'est, & où l'on fait voir les mauvaises conséquences des explications trop subtiles que l'on en donne* (Amsterdam, 1715).

[19] See previous note. Coste translated Locke's *Some Thoughts concerning Education* (1695) and the *Essay* (1700). On Coste and Locke see John Milton, 'Pierre Coste, John Locke and the Third Earl of Shaftesbury' in Sarah Hutton and Paul Schuurman (eds), *Studies on Locke: Sources, Contemporaries and Legacy* (Dordrecht: Springer, 2008), 195–223 and the literature cited therein. Also Delphine Soulard, 'Les Huguenots du Refuge et la diffusion de la pensée politique de John Locke auprès du public francophone', in Ann Thomson, Simon Burrows and Edmond Dziembowski (eds), *Cultural Transfers: France and Britain in the Long Eighteenth Century* (Oxford: SVEC, 2010), 147–59.

Lady Masham very likely also had some correspondence with John Norris: she certainly knew Norris, but broke off relations after he mistakenly announced in print that she had gone blind. If they did exchange letters, these letters do not survive.[20]

10.3 Latitudinarianism and Toleration

Locke testifies to Lady Masham's interest in and knowledge of religious as well as philosophical matters, and he thought very highly of her abilities.[21] Further testimony on this score comes from an anonymous correspondent who complimented her by referring to her 'happy genius', which 'leads you to enquire into all kinds of Learning, and especially to judg of Theologicall Points, therein shewing yourself to be the True Offspring of so Learned and Judicious a Parent'.[22]

Although, Lady Masham quotes from the Bible rarely in her letters, her publications show that she certainly knew her Bible. Someone who appreciated her opinion on Scripture was Isaac Newton who discussed his interpretation of the Book of Daniel on a visit to Oates.[23] Her theological reading included Stillingfleet's sermons and *Fanaticism fanatically imputed to the Catholick Church* (1672), Le Clerc's *Ars critica* (1696) and Jacquelot's *Examen de la théologie de M. Bayle* (1706). Evidence of her technical knowledge of theology is to be found in a letter which she wrote to Jean Le Clerc in May 1704 where she mentions different views on the Trinity, noting differences between 'Nominal, and Real Trinitarianisme' and between either of them and Socinianism.[24] This is one of the very few places where Damaris Masham delivers an opinion on a matter of doctrine (discussed below).

Although there is no discussion of religious doctrine in her letters to Locke, they do discuss religious philosophy. Examples are her discussion of the *Select Discourses* of John Smith at an early point of her correspondence with Locke, in which she discusses the power of reason to attain knowledge of God.[25] Another point where philosophy and religion overlap is the letter in which she comments on Locke's

---

[20] For the Shaftesbury letters: London, National Archives, PRO 30/24/20, fols 266–7 & 273–4. Norris dedicated to her his *Reflections on the Conduct of Human life* (1690). However, relations between them soured because in the Dedication he erroneously announced that she had gone blind, without checking his facts first. This still rankled with Damaris Masham in 1704, when she recalled it in a letter to Le Clerc. *Epistolario*, 2: 391, letter 342.

[21] Locke, *Correspondence*, 4: 237, letter 1375.

[22] Ibid., 5:601, letter 2063.

[23] Ibid., 3: 288, letter 1405.

[24] Le Clerc, *Epistolario*, 2: 445–7, letter 380.

[25] Locke, *Correspondence*, 2: 485, 488, 500, letters 684, 687, 696.

proof of the existence of God in an un-named book which was probably the *Abrégée*.[26] Her correspondence with Leibniz is the most philosophical of her epistolary exchanges, but even here, the topics have a religious dimension—the nature of God and Providence. Another place where religion and philosophy overlap is Lady Masham's defence of her father's philosophy against the charge of atheism, which she does in letters to Leibniz and Le Clerc. Her philosophical treatment of religious topics shows that she held not only the compatibility of faith and reason, but also, that reason has a major role in religious matters, which she confirms when she tells Le Clerc that she agrees that faith is compatible with reason and philosophy. Theologically this aligns her with the Cambridge Platonists and their latitudinarian successors. There is further circumstantial evidence for Lady Masham having strong latitudinarian sympathies: before she married, Damaris Cudworth had close personal contacts with Latitudinarian circles—she knew the wife of Edward Stillingfleet ('Urgunda' of the letters) and, in letters to Le Clerc and Van Limborch she recalls being present, as a young woman, at the home of Stillingfleet when he was Dean of St Paul's.[27] She was married in Stillingfleet's church. On both Locke's and her own testimony she was a firm believer in religious toleration. She told Jean Le Clerc that she believed 'Liberty of Conscience is the unquestionable Right of Mankind'. [28]

Lady Masham was also well aware of religious differences, not just between Protestants and Catholics, but among Protestants, and also within the Church of England. In 1705, she told Van Limborch that she believed that many ministers of the Church of England were anti-Calvinists who, like her father, held Arminian beliefs that were not compatible with the Calvinistic articles of the Anglican Church. But they all seemed to have found a way of squaring this with their consciences. By what means, she didn't know ('la plupart … trouve je ne sai comment, les moyens de satisfaire leur conscience en declarant qu'ils donnent à ces articles non seulement *consensum* mais aussi *assensum*'). When she was still a young woman he had quizzed her father about the discrepancy between his theological views and the articles of the Anglican faith to which he conformed at the Restoration. She adds that since this applies to so many excellent men ('tant d'excellens hommes'), she dare not censure the practice ('qu'à peine osé-je censurer cette practice').[29]

[26] Ibid., 3: 434, letter 1040.

[27] Letter to Van Limborch, 17 September 1705, printed in Hoeven, *De Joanne Clerico*, 137, Le Clerc, *Epistolario*, 2: 497–517, letter 380.

[28] Ibid.

[29] Letter to Van Limborch, 17 September 1705, printed in Hoeven, *De Joanne Clerico*, 137. This letter contains important evidence that Ralph Cudworth originally held Calvinist views, and that he did not repudiate them until after he was elected fellow of Emmanuel College. Cudworth himself had told Van Limborch that members of the Restoration Church of England represented a broad spectrum of beliefs, from the Calvinistial to Socinian. Cudworth to Limborch, 16 March 1675, Amsterdam, Universiteits bibliotheek, MS M.21.c.

10.4 Disputes and Differences

Lady Masham was well aware, then, of differences of theological stance, even that her father was a closet dissenter (or 'heretic' as one of his admirers jokingly called him). It was perhaps in the light of this comprehensive view of doctrinal matters that she expressed disinclination to enter into controversy about religious beliefs. She told Locke that she did not 'intend to set up a Lecture of Divinitie, Or Enter into All the Disputes and Differences of Religion', since matters of contention are difficulties created by men.

> I think that I may much more Advantageously employ my Houres in Pursuing the End of these speculations then in indeavouring to Extricate those Difficulties that the Witts of Men have Intangled them with, Which being Needless to my self, can be no Part of my Obligations.[30]

By 'Needless' she means that it is useless or irrelevant because disputes about doctrine not only do not achieve the purpose of strengthening religious faith, but actual obstruct it. She told Le Clerc that she regarded religious disputes as useless and divisive, leading to baseless charges of heresy over opinions which have no basis in Christian doctrine:

> It is really sad to see that whilst such Teachers of the People do with so cruel heart as they too often do, fall upon some men only for dissenting from the in opinions which are no Doctrines of Christianitie, and represent for dangerous Hereticks such (for example) as you are.[31]

Although she disliked controversy, Damaris Masham was, nevertheless, fully prepared to voice objections strenuously, if occasion required—as when she complained to John Covel, Master of Christ's College, for having licensed a book which denigrated her household as 'the seraglio at Oates'.[32] And she did not eschew debates about religion altogether. Her first excursus into print, after all, was a critique of the Malebranchian views of John Norris. Her *A Discourse Concerning the Love of God* (1696) targets John Norris's Malebranchian account of the relationship of God to the world, which is the subject of *Letters concerning the Love of God* (1695) which prints John Norris's exchange of letters with Mary Astell.[33] In this Norris defended

---

[30] Locke, *Correspondence*, 3: 432, letter 1040.

[31] Le Clerc, *Epistolario*, 2: 445–7, letter 364.

[32] Ibid., 6: 218, 252, letters 2323, 2347 for Covel's apology for signing the imprimatur.

[33] Modern edition: Mary Astell and John Norris, *Letters Concerning the Love of God*, ed. E. Derek Taylor and Melvyn New (Aldershot: Ashgate, 2005). I have argued that Lady Masham was also targeting Mary Astell's arguments. For a balanced riposte, see Jacqueline Broad, 'Adversaries of Allies? Occasional Thoughts on the Masham-Astell Exchange', *Eighteenth Century Thought* 1 (2003): 123–49. Also, William Kolbrener and Michal Michelson (eds), *Mary Astell: Reason, Gender, Faith* (Aldershot: Ashgate, 2007). The Masham attack on Malebranchianism probably had its origins in conversations with Locke at Oates. Locke too, drafted a critique of Malebranche. See Paul Schuurman, 'Vision in God and Thinking Matter Locke's epistemological agnosticism used against Malebranche and Stillingfleet', in Hutton and Schuurman, *Studies on Locke*, 177–93.

his idea of 'seeing all things in God', and his claim that we are not motivated to love God by our experience of the physical word. Lady Masham objected that Norris's Malebranchian understanding of our relationhip to God undermines the very basis of religious belief. The controversial character of *A Discourse*, belies her claims in her letters that she did not wish to be drawn into 'Disputes and Differences of Religion'. Arguably Lady Masham made a distinction between broadly philosophical debate and disputes about theological doctrine. Although fundamentals of religion are concerned in her critique of Norris (the very basis of religious belief) she did not pick on points of doctrine or creed. As the title of the French translation indicates, this work is, as much as anything about the proper nature of rational discussion of religion and the danger of overly sophisticated arguments—*où l'on explique ce que c'est, & où l'on fait voir les mauvaises conséquences des explications trop subtiles que l'on en donne.*

10.5 Philosophy, Religion and Atheism

Another religio-philosophical debate in which Lady Masham was involved, concerned the philosophy of her father, Ralph Cudworth, whom Pierre Bayle charged with atheism. This was a quarrel between Bayle and Le Clerc. Bayle was responding to the selections from Cudworth's *True Intellectual System* which Jean Le Clerc had published in French translation in *Bibliothèque choisie*.[34] The controversy arose from Bayle's critique of Cudworth's hypothesis of Plastic Nature. It was not a religious controversy in the sense of a dispute about doctrine, but it was religious in the sense that what was at stake was the existence of God and the nature of providence—and, of course Ralph Cudworth's good name as a *religious* philosopher. This controversy has been discussed by Rosalie Colie (1957), so I shall confine myself to some observations about the nature and priorities of Lady Masham's involvement which was primarily through letter exchange.

Lady Masham's part in the debate illustrates the constraints which affected her entering into a public dispute. She did not intervene directly with Bayle—at least, not publicly. Rather, she used Le Clerc as an intermediary, apparently sending him objections for him to use in order to reply to Bayle in *Bibliothèque choisie*.[35] One such is mentioned explicitly by Le Clerc, but rather than print it in *Bibliothèque choisie*, he forwarded it to Bayle.[36] The fortunes of the letter illustrates that she was very much dependent on the goodwill of her contacts. Initially, Pierre Coste removed it from a package which he had been instructed by her to send to Le Clerc. Coste told Le Clerc

[34] Bayle's critique of Cudworth originally appeared in Basnage de Beauval's *Histoire des Ouvrages des Savants* and his own *Continuation des pensées diverses sur la comète*, republished in his *Œuvres Diverses* (1727).

[35] *Bibliothèque choisie*, 5, item 4 (1705) and 6, item 13 (1705).

[36] Ibid., 6, item 13 (1705).

that he had done so because he thought that Bayle was going to reply to her complaints.[37] Coste then changed his mind and sent the letter on to Le Clerc, who forwarded it on to his adversary. Bayle, in his turn, did not reply directly to Lady Masham, but responded in *Histoire des Ouvrages des Savans*.[38] He agreed that Lady Masham 'se plaint avec raison de mon procéde à l'égard de Mr. Cudworth son pere', laying the blame on Le Clerc's translation (Colie 1957).[39] In this spat, Lady Masham did not respond directly to Bayle. Instead she sought to involve others, notably Leibniz with whom she raised the matter of Bayle's attack on Cudworth, probably hoping to draw him in on Cudworth's side. If her indirection highlights some of the hesitancy Damaris Masham felt about public intervention in debate, it also suggests uncertainty on the part of both Coste and Le Clerc, about publishing her views.

The Cudworth-Bayle controversy was not a matter of doctrine and did not explicitly revolve around Damaris Masham's own views either philosophically or theologically. Although she was evidently in sympathy with her father's philosophy and theology, she doesn't present herself as his disciple. In defending his namc as a philosopher of religion, she in effect makes herself an apologist for religion in the same mould as he was. The Bayle-Le Clerc spat both made her father's philosophy more widely known and helped to bring her to the attention of the *savans de l'Europe*. Barbeyrac, for example regarded her as 'Savante d'un mérite distingué et Fille du Docteur CUDWORTH'.[40] This was, of course, secular fame as a lady of letters, not as a model of piety.

10.6 Trinitarianism

A theological controversy of which she was certainly aware was the controversy variously known as the Trinitarian Controversy or the Unitarian Controversy, which raged in England in the 1690s, during the time of Locke's residence at her home (Champion 1992; Marshall 1994; Nuovo 2000; Dixon 2003; Hedley 2005). Sparked by Stephen Nye's *Brief History of the Unitarians* (1689) the Unitarian controversy was fuelled by differences of opinion between the Latitudinarian and High Church parties within the Church of England: Nye and the leading contributors William Sherlock, Robert South and Arthur Bury were all Anglican clergymen.[41] It was against the background of the Unitarian controversy that Locke became embroiled

[37] Le Clerc, *Epistolario*, 3: 527–8, letter 3383.

[38] 'Reflexions de Mr. Bayle sur l'Article VII du 6 Tome de la Bibliothèque choisie de Mr. le Clerc', *Histoire des Ouvrages des Savans*, Dec. 1704., art XII, 540–4.

[39] Le Clerc, *Epistolario*, 2: 496–7, letter 364.

[40] *Éloge* of Locke by Barbeyrac printed in Le Clerc, *Epistolario*, 6: 498.

[41] Nye, too, though a Unitarian, was a minister of the Church of England. The anti-trinitarian sympathies of some of the Anglican participants in this debate bear out Cudworth's claim to Philip Van Limborch in 1675 that the Church of England harboured Socinians within its ranks. Cudworth to Limborch, 16 March 1675, Amsterdam, Universiteits bibliotheek, MS M.21.c.

in his well-known debate over the Trinity with Edward Stillingfleet. And this took place during the period of his closest contact with Damaris Masham, but also the period when they no longer needed to write to one another. For this reason, the extent of her direct involvement in it is not immediately clear, and it must be pieced together from both circumstantial evidence and correspondence with others.

Lady Masham had a link to the Trinitarian/Unitarian debate independently of Locke, through her father. One of the reasons for a hostile reception of Ralph Cudworth's *True Intellectual System* was his discussion of the doctrine of the Trinity, and this makes him a key reference point in the Unitarian Controversy, in which he was invoked by opposing parties (Hedley 2005; also Carter 2011). The first attack on Cudworth's Trinitarianism antedates the controversy proper. It was by John Turner, who prefaced his *Discourse concerning the Messias* (1685) with 'a large preface, asserting…the doctrine of the Blessed Trinity against the late writer of the Intellectual System', i.e. Ralph Cudworth. Subsequently Cudworth figures on both sides of the Trinitarian controversy: in his *Considerations on the Explications of the Doctrine of the Trinity* (1693), Stephen Nye commends Cudworth's 'excellent labours' in his discussion of the Trinity. On the opposite side of the debate, Cudworth was attacked by William Sherlock in his *Vindication of the Doctrine of the Holy and Ever Blessed Trinity* (1690), for claiming that the early fathers were Tritheists. Sherlock's opponent, Robert South identifies Cudworth as the source for Sherlock's Platonising account of the persons of the Trinity.[42] Although Locke did not become embroiled in this dispute, his correspondence with Benjamin Furly shows that he followed it. Locke, too, regarded Cudworth as a source for 'the Original of the Trinitarian doctrines, from whom they are derived … by whom they were invented', and noted in particular, Cudworth's useful 'parallelism betwixt the Ancient or Genuine Platonick, and the Christian Trinity'.[43]

Absence of letters between Locke and Lady Masham during this period make it difficult to be certain whether she and Locke discussed this dispute, or whether she had any part in his debates with others. But there are good grounds for believing that she was not only *au courant* with the debates concerning the Trinity, and that she was more than just a 'reader over the shoulder' of Locke. Lady Masham's close personal contacts with Stillingfleet, means that she must have been aware of the Stillingfleet-Locke debate, and her view of it was no doubt complicated by her personal acquaintance with Stillingfleet. It is more than likely that she was familiar with at least some of Locke's correspondence at this time (a remark she makes to Le Clerc

[42] *Tritheism charged upon Dr. Sherlock's new notion of the Trinity*, (1695). Sherlock's 'Evangelist *Plato*', he says, is based on Cudworth's account of the Platonic Trinity. This was not lost on Joshua Toulmin in his *An Historical View of the State of the Protestant Dissenters in England* (1814) who noted similarities in Sherlock's and Cudworth's views, 'concerning the three divine persons … they both apprehended the three persons to be as distinct and different, and as really three several intelligent beings and substances'. *John Locke and Christianity*, ed. Victor Nuovo (Bristol: Thoemmes Press, 1997), 112.

[43] 'Adversaria theologica'. Quoted in John Marshall, *John Locke, Resistance, Religion and Responsibility* (Cambridge: Cambridge University Press, 1994), 395.

about his having 'complained of your want of leisure wherein to write to Mr Locke',[44] indicates that she was aware of those letters of his in which he had done so).

Evidence that Lady Masham was aware of these controversies surrounding the Trinity comes from the already cited anonymous letter regarding John Edwards' book *Socinianism Unmask'd*, which Lady Masham received in 1696.[45] The occasion of this letter was John Edwards's attack on the recent but anonymous publication of Locke's *The Reasonableness of Christianity* (1695). In his *Socinianism Unmask'd* Edwards accused the author of *The Reasonableness* of Socinianism. Why the anonymous writer of this letter should have written to Lady Masham to alert her to this attack is unknown—though, Edwards evidently respected Cudworth's scholarship, to judge by his *Some thoughts concerning the several causes and occasions of atheism* (1695) (which also includes an attack on Locke's *The Reasonableness of Christianity*).[46] This might indicate that the letter-writer's intentions were friendly; if so, his choosing to remain anonymous is puzzling. But it is not at all clear from the letter whether either Edwards or the letter-writer knew that Lady Masham had close links to Locke, or that they suspected that Locke was the author of *The Reasonableness*. If these connections were known, the purpose of sending her the letter might have been hostile, intended as a way of smoking out the identity of the author of the *The Reasonableness*. That itself indicates that he had no personal acquaintance with Damaris. Whatever the case, Edwards's attacks prompted Locke to reply (anonymously) in print with his *A Vindication of the Reasonableness of Christianity from Mr. Edwards's Reflections* (1695).[47]

Further evidence that Lady Masham was *au courant* with the furore surrounding the Trinity is to be found in her extant letters to Jean Le Clerc. The most important letter is one previously cited, which she wrote to Jean Le Clerc in May 1704. In this she broached the topic of the Trinity, making direct reference to current disputes ('Doctors, who quarrel one with another upon this Controversie'). The occasion was her gift of George Bull's *Opera omnia* to Le Clerc as a token of gratitude for the translation of Ralph Cudworth which he had sent her. Bull was an authority on the Trinity, having defended it in his *Defensio fidei Nicenae* (1685). Lady Masham's comments on the Trinity are her most detailed personal views on any religious doctrine in her letters.

> Nominal, and Real Trinitarianisme, differ methinks, at lest as much one from another as socianisme from either of them. But (God be thank'd) it suffices (here) on all hands, for the orthodoxie of those of vulgar understanding but to Beleeve or say that they Beleeve a

[44] Lady Masham to Jean Le Clerc, Le Clerc *Epistolario*, 2: 389, letter 342.

[45] The book in question was John Edwards. *Socinianism unmask'd a discourse shewing the unreasonableness of a late writer's opinion concerning the necessity of only one article of Christian faith, and of his other assertions in his late book, entituled, The reasonableness of Christianity as deliver'd in the Scriptures, and in his vindication of it* (1696).

[46] The anonymous letter-writer tells her that Edwards is 'a great Admirer and Honourer of your Ladyship' (Locke, *Correspondence*, 5: 601, letter 2063).

[47] Edwards's attack on Locke is mentioned in the Van Limborch-Locke correspondence in 1697: Locke, *Correspondence*, 6: 43 and 111, letters 2222 and 2256. There follows a long exchange about the unity of God and divinity of the Messiah, ibid., letters 2318, 1240, 2352, 2396, 2410.

> Trinitie; no Matter of what. And the Doctors, who quarrel one with another upon this Controversie in theire Learned Writeings, instruct us Out of the Pulpit onely to hold fast the Form of sound words. And tell us but what we should Not Beleeve; not what we should Beleeve, in a matter on which (whatever they think in regard of themselves) I from hence Hope they are persuaded that Our salvation dos not depend. Such as I am being thus safe in Ignorance, whilst they can no sooner explicitly Beleeve, than they become Heterodox on one side, or Other. We should, I think, be in the right not to Perplex ourselves with this Question, had we not yet a better Reason for not doing so, from the silence of the Scripture herein. To distinguish Nominal Trinitarianisme from Socinianisme; and Real Trinitarianisme from Tritheisme; and to Decrie which of these Opinions is best supported by Antiquitie, are matters for yor learned and knowing men.[48]

On the evidence of this extract, Damaris Masham was certainly conversant with theological technicalities. It also shows her scepticism about the value of theological disputes, echoing her earlier observations to Locke about the 'Difficulties that the Witts of Men have Intangled them with'.[49] Disputations of this sort are an academic matter for the learned, and not a way to arrive at judgements about faith and are not binding in point of faith. But this letter is revealing of more about her stance on disputes and her own personal beliefs. The unattributed quotation from 2 Timothy 13, 'to hold fast the Form of sound words', invokes the authority of St Paul for her view that it is enough for ordinary Christians ('those of vulgar understanding') simply to state their belief in the Trinity ('to Beleeve or say that they Beleeve a Trinitie') without stating exactly what that entails ('that they Beleeve a Trinitie; no Matter of what'). Her rejection of disputation as a means of solving creedal differences is itself a tolerationist position. But this letter indicates that her latitude in point of Trinitarian doctrine is underpinned by a measure of agnosticism about the doctrine itself. The letter is also revealing for what she says about the appropriate *conduct* of the Christian within the Church, that quarrels of this kind go against the Christian spirit, a position which recalls that of her father in his great House of Commons sermon of 1647.

Lady Masham's choice of gift for Le Clerc was dictated by his theological preferences as she perceived them. She chose it, she says, because she had noticed he had expressed positive views about George Bull in his *Ars critica*. Since Bull was no friend of the Platonic trinity defended by Cudworth, Lady Masham clearly hadn't made her choice of gift on Cudworthian grounds. And his *Opera omnia*, would of course, count as a work by one of the 'learned heads' to whom she refers in her letter. But Lady Masham is unlikely to have made a gift of a book of which she did not approve, and Bull did have links to leading latitudinarians (Tillotson and Burnet). On the matter of the Trinity, Bull was not neutral, at least not in the context of the Trinitarian disputes of the time. He had published his *Defensio fidei Nicenae* (1685) in the teeth of opposition from Oxford clergy, like Thomas Barlow, who denounced him as a Socinian. The appeal of Bull's position to more latitudinarian Churchmen would seem to have been that what he proposed was 'soft' or 'blurred'

---

[48] Lady Masham to Jean Le Clerc, Le Clerc, *Epistolario*, 2: 445–7, letter 380.

[49] Locke, *Correspondence*, 3: 432, letter 1040.

Trinitarianism (if one may call it that). This seems to have been the view of Le Clerc, who regarded Bull's account of the Trinity as equivocal. His defence of the Nicene doctrine was nevertheless open to an Arian interpretation in the manner of Samuel Clark.[50] Lady Masham's repudiation of distinctions between types of Trinitarianism ('Real'/'Nominal'/'Tritheism') certainly fits with a 'blurred' position on the Trinity, and her non-hostile reference to Socinianism alongside Trinitarians of different stripes suggests a very wide degree of tolerance on her part towards differing theological views. More radically, however, her view on the Trinity seems to be underscored by doubts about the doctrine itself, for she remarks that a precise position on the Trinity is irrelevant if the doctrine is not grounded in scripture: as she said to Le Clerc 'the silence of the Scripture', makes it 'a matter on which … Our salvation dos not depend'.

The third aspect of this letter which I want to highlight is Lady Masham's comment on the relevance of theological disputes to ordinary Christians ('those of vulgar understanding'). She tells Le Clerc that she believes that it is possible to be an orthodox Christian, without a deep understanding of the complexities of doctrine. Being 'safe in Ignorance' is the condition of the generality of Christians. And this is the position which, she claims, the practising clergy encourage their flocks to adopt. The clergy 'instruct us Out of the Pulpit onely to hold fast the Form of sound words. And tell us but what we should Not Beleeve; not what we should Beleeve'. It is perhaps surprising that someone who set so much store by reason in religious matters should recommend assent without understanding, acceptance of the mere 'Form of sound words'. However, it is a practical approach to dealing with differences, and one which, in the spirit of St Paul, is conducive to genuine faith. By contrast, through their disputations theologians risk arguing themselves into heterodoxy. Her comments make the common sense point that ordinary Christians do not, in practice, concern themselves with the finer points of doctrine. She does not in fact state that they should *not* do so. Rather, like John Smith and Ralph Cudworth before her, she did not regard understanding as a pre-requisite for salvation. As Ralph Cudworth told the House of Commons, 'no man shall ever be kept out of heaven, for not comprehending mysteries that were beyond the reach of his shallow understanding'.[51] There is another parallel with Ralph Cudworth (as recalled in her letter to Van Limborch) in the way she draws a distinction between outward conformity and inward dissent—the position of someone prepared to accept outward conformity, or radical indifference, for the sake of ecclesiastical peace, provided inner conscience is left free. Indeed, as we have seen, from her letter to Van Limborch, she thought that most Church of England clergymen made accommodations of this kind between their own (arminian) beliefs and the 39 articles of the Anglican creed.

---

[50] 'Il semble que Mr *Sam. Clarke*, et plusieurs autres Theologiens Anglois donnent dans un Arianisme mitigé, qu'il[s] prétendent avoir été le sentiment des PP. des 3 premiers Siecles, qui se sont en effet expliquez là-dessus d'une manière assez équivoque, comme il paroît assez par le P. *Petau* et par Mr. *Bull*, qui se donne la torture, pour leur donner un bon sens'. Le Clerc to Turretini, 1723, *Epistolario*, letter 232.

[51] Ralph Cudworth, *A Sermon Preached before the House of Commons* (Cambridge, 1647), 14.

Radical views like Lady Masham's on the Trinity (which amount to what might be called Trinitarian agnosticism) and her tolerant attitude to religious differences, even on such a central doctrine as the Trinity, are also consistent with Locke's theological position. And, like Locke, she would have had to be careful about expressing any theological doubts publicly, especially views on such an explosive issue as the Trinity. With views like these, she was very likely to have been party to ecclesiological debates at Oates during Locke's residency, and to have discussed topical disputes, including Locke's debate with Stillingfleet. But there is no reason to assume that it was only as a result of Locke's presence that she took an interest. On the contrary, as we have already seen, there are Cudworthian roots for the position she takes in her letter to Le Clerc which echoes the eirenicism of her father both in the wide latitude of her religious tolerance, but also in her adoption of a stance of outward conformity masking inner dissent. Here Damaris Masham certainly proved herself to the True Offspring of so Learned and Judicious a Parent'.[52] But she was her own woman, with her own views, which she debated through the medium of private letters.

References

Carter, Benjamin. 2011. *"The little commonwealth of man": The Trinitarian origins of the ethical and political philosophy of Ralph Cudworth.* Leeuven: Peeters.

Champion, Justin. 1992. *The pillars of priestcraft shaken.* Cambridge: Cambridge University Press.

Colie, Rosalie. 1957. *Light and enlightenment.* Cambridge: Cambridge University Press.

Dixon, Philip. 2003. *Nice and hot disputes: The doctrine of the trinity in the seventeenth century.* London: T&T Clark.

Goldie, Mark. 2004. *John Locke and the Mashams at Oates.* Cambridge: Churchill College.

Harris, Frances. 2003. *The transformations of love: The friendship of John Evelyn and Margaret Godolphin.* Oxford: Oxford University Press.

Hedley, Douglas. 2005. Persons of substance and the Cambridge connections some roots and ramifications of the Trinitarian controversy in seventeenth-century England. In *Socinianism and Arminianism. Antitrinitarians, Calvinists and cultural exchange in seventeenth-century Europe*, ed. Martin Mulsow and Jan Rohls, 225–40. Leiden: Brill.

Hutton, Sarah. 1993. Damaris cudworth, Lady Masham: Between Platonism and enlightenment. *British Journal for the History of Philosophy* 1: 29–54.

Hutton, Sarah. Forthcoming. Religion and sociability in the correspondence of Damaris Masham (1658–1708). In *Women and religion in England*, ed. Sarah Apetrei and Hannah Smith. Aldershot: Ashgate.

Marshall, John. 1994. *John Locke, resistance, religion and responsibility.* Cambridge: Cambridge University Press.

Nuovo, Victor. 2000. Locke's theology, 1694–1794. In *English philosophy in the age of Locke*, ed. M.A. Stewart, 183–215. Oxford: Clarendon Press.

Phemister, Pauline. 2004. "All the time and everywhere, everything's the same as here": the principle of uniformity in Leibniz's correspondence with Lady Masham. In *Leibniz and his correspondents*, ed. Lodge Paul, 193–213. Cambridge: Cambridge University Press.

[52] In the words of the anonymous letter-writer quoted above at note 21.

Phemister, Pauline, and Justin Smith. 2007. Leibniz and the Cambridge Platonists and the debate over plastic natures. In *Leibniz's philosophy and the English-speaking world*, ed. Phemister Pauline and Brown Stuart, 95–110. Dordrecht: Springer.

Simonutti, Luisa. 1987. Damaris Cudworth Masham: una Lady della Repubblica delle Lettere. *Scritti in Onore di Eugenio Garin*. Pisa: Scuola Normale Superiore, 141–65.

Simonutti, Luisa. 2008. Circles of *virtuosi* and "charity under different opinions": The crucible of Locke's last writings. In *Studies on Locke: Sources, contemporaries and legacy*, ed. Hutton Sarah and Schuurman Paul, 159–76. Dordrecht: Springer.

Sleigh, Robert. 2005. Reflections on the Masham correspondence. In *Early modern philosophy: Mind, matter, and metaphysics*, ed. Mercer Christia and O'Neill Eileen, 119–27. Oxford: Oxford University Press.

Chapter 11
Evangelical Calvinists Versus the Hutcheson Circle: Debating the Faith in Scotland, 1738–1739

James Moore

Early in the year 1738, an open letter was addressed to Francis Hutcheson, Professor of Moral Philosophy, in the University of Glasgow. In that letter, Hutcheson was accused by a former student of teaching 'gross and dangerous errors' to the moral detriment of his students. He was also reminded that he had taken an oath on his appointment that he would teach nothing contrary to the Westminster Confession of Faith. This letter provoked a 'vindication' of his teaching 'from the calumnious aspersions of a late pamphlet' by Hutcheson's students and former students. Later in the same year, another open letter was published. It was addressed to 'the Reverend and Renowned Mr. Ebenezer Erskine' the leader of the evangelical Calvinists in Scotland at that time. The title page of this letter identified the author as Euzelus Philalethes (zealous lover of truth), the pseudonym employed by the writer of the first letter. The second letter was, however, a hoax. It was a satire on Hutcheson's student critic and on evangelical Calvinists in general. In October 1738, a sermon was delivered in the Synod of Glasgow and Ayr by a minister, described later in the century, as a 'bosom friend' of Francis Hutcheson. A private letter from Hutcheson to another friend reveals that Hutcheson himself contributed matter for this exercise. The sermon was published in Edinburgh in 1739.

This was a moment of some significance in the Scottish enlightenment, when the philosophy of Francis Hutcheson was defended in a pamphlet, an open letter and a sermon from an attack inspired by the doctrines of evangelical Calvinism. It is an episode that has remained largely unnoticed, however; perhaps because such an explicit opposition between Reformed orthodoxy and what we have come to call the enlightenment in Scotland was not typical of the style of enlightened thinkers, professors and ministers of the Church of Scotland.

J. Moore (✉)
Department of Political Science H-1225-28, Concordia University,
De Maisonneuve Blvd, 1455, H3G1M8 Montreal, QC, Canada
e-mail: jmoore@netrover.com

A. Dunan-Page and C. Prunier (eds.), *Debating the Faith: Religion and Letter Writing in Great Britain, 1550-1800*, International Archives of the History of Ideas 209,
DOI 10.1007/978-94-007-5216-0_11, © Springer Science+Business Media Dordrecht 2013

In recent scholarship on the Scottish enlightenment, a sharp contrast has been drawn between the explicit challenges to Calvinist orthodoxy that occurred, early in the eighteenth century, in Geneva, northern Ireland, England and America with the 'silent', 'almost invisible' enlightenment in Scotland: 'Curiously, the Church which contributed so much in the way both of ideas and personnel to the flowering of the Scottish enlightenment did not witness a sustained assault on the authority of the Westminster Confession of Faith. The mainstream Scottish enlightenment was—at least superficially—a Calvinist affair.' (Kidd 2000, 32–3). This account of the enlightenment in Scotland is consistent with Richard Sher's description of the Moderate Literati of Edinburgh: 'None of the Moderates in the William Robertson circle had any scruples about subscribing to the Church's rigorously Calvinist creed, the Westminster Confession of Faith, and none of them ever overtly denied its fundamental tenets' (Sher 1985, 35). And the same reluctance to debate the terms of the Confession of Faith characterized the writings and the sermons of William Wishart, Robert Wallace and Patrick Cuming earlier in the century (Sefton 1983, 187–8).

It should be added, however, that all of these clergymen, philosophers and men of letters preached, taught and wrote in or near Edinburgh for most of their careers. And all of them received their education in the study of divinity at the University of Edinburgh where a style of instruction had been put in place by the Professor of Divinity, William Hamilton, who held this position from 1709 to 1732. It was his practice to remain silent on controversial principles of theology: the Incarnation of God in Christ, the Atonement made by Christ for our sins, the Trinity. This style was continued by his students and successors in their preaching and writing.

The same scholars who have described the style of the Edinburgh Moderates have recognized, however, that a different academic culture prevailed in Glasgow, where the Professor of Divinity, the Reverend John Simson, was charged with teaching Arminianism (1716–17) and later the Arian heresy (1726–29) (Sefton 1983, 187–8). Richard Sher excluded Francis Hutcheson from the ranks of the Moderate Literati, while remarking the influence of Hutcheson's brand of Stoicism for their understanding of Christian morality (Sher 1985, 16–7, 325). And the dispute in northern Ireland over the requirement that ministers subscribe the *Confession of Faith* 'infected the academic politics of the University of Glasgow, a notorious seminary of heresy, hotheadedness and rebellion' (Kidd 2000, 48).

11.1 Shaftsbury's Ghost conjur'd

Even before his arrival in Glasgow, in 1730, to take up the position of Professor of Moral Philosophy, apprehensions were expressed by strict or orthodox Presbyterians about Hutcheson's religious principles. Robert Wodrow, a clergyman, historian and diarist, wrote, following Hutcheson's election: 'Hou the principles he goes on agree with the truths generally received in this Church, and what influence his teaching them here may have, time will discover.' (December 1729). And after Hutcheson's inaugural lecture, a year later, he recorded that Hutcheson 'delivered it soft and low and it was not well understood; his character and carriage seem prudent and cautious, and this will be the best vidimus of him'.[1]

A letter has come to light that was written in 1732 by a student of Hutcheson's who hoped that his professor would agree with him in his religious convictions, and 'believe me, I would rather gain your Approbation than have Millions of Worlds for my Admires'[sic?] He proceeded to offer his opinion that 'Christ came into the World a person, if not the son of God, Yet of a Nature superior to Ours ... Mankind has reaped a great many blessings from him, he having if not redeemed them from Death by the price of his blood, at least set before them an Example of a Holy Life and laid down the precepts of a Divine Morality'.[2] His student was hoping that Hutcheson would agree with him that there was no need to call Christ the son of God; he was undoubtedly a superior person and he provided an example of a holy life. It was unnecessary to say that he has redeemed us by the sacrifice of his blood. This student was an Arian; he did not believe in the divinity of Christ, or the Incarnation, or the Atonement for our sins made by Christ. We do not have Hutcheson's reply to this letter; perhaps he never wrote a reply; opinions as heterodox, as remote from Reformed dogmatism as these would be better confined to conversation.

Hutcheson had another student at the University of Glasgow who did not approve of his teaching and published an open letter to him to that effect in the year 1738: *Shaftsbury's Ghost conjur'd; Or, A Letter to Mr. Francis Hutcheson, Professor of Moral Philosophy in the University of Glasgow. Wherein Several gross and dangerous Errors, vented by him in the Course of his Teaching, are brought to light, and refuted* is a small tract of over 40 printed pages. The author of this letter was Hugh Heugh, the eldest son of a Presbyterian minister, the late Reverend John Heugh (1688–1731), Minister of the Church of Scotland in Kingoldrum.[3] The Reverend John Heugh was deeply opposed to 'enemies of the church' such as the Reverend John Simson. He thought that Arians were guilty of heresy in 'denying Christ to be the true and independent God and Making God a mere Creature' (MacGill 1852, 9). It was his dying wish that the Reverend Alexander Moncrieff of Culfargie (1695–1761) would act as guardian of his son, Hugh (MacGill 1852, 9, 13). Moncrieff would support the Reverend Ebenezer Erskine (1680–1754) in his protest against an Act of the General Assembly of the Church of Scotland (1732) that would have abolished patronage but confined the calling of ministers to heritors and elders (Lachman 2004). Moncrieff was sharply critical of 'the Scheme of Selfish-Love,

---

[1] Robert Wodrow, *Analecta, or a History of Remarkable Providences* (Edinburgh, 1854), vol. 4, 99, 187.

[2] National Library of Scotland (thereafter NLS) MS 9252, 80–1, Letter from Patrick Lang, September, 1732. The student was careful not to implicate his professor in his own heterodox religious opinions, but he hoped and seems to have expected that his professor would agree with him.

[3] Hamilton Macgill, *The Life of Hugh Heugh*, D.D. (Edinburgh: A. Fullarton and Co., 1850), 14. The subject of this biography, Dr. Hugh Heugh (1782–1846) was a minister in the United Secession Church and the son of John Heugh (1731–1810) who was a brother of Hutcheson's critic. They had three sisters; all married ministers who seceded from the Church of Scotland.

laid down by Mr. Archibald Campbell' in his moral philosophy; but this did not inhibit him from criticizing other philosophers who 'may amuse themselves with the Notion of God's Benevolence…'[4] Hugh Heugh had attended Hutcheson's class in moral philosophy in 1734–35,[5] and he was impressed by Hutcheson's lectures when he first heard them: 'my Fancy was tickled with your chimerical Ideas of Virtue and a Moral Sense'. But then Heugh returned to the class a second time.

> I made it my business, when I heard any Thing that I was not fully satisfied about, to write it down as soon as I returned to my Room; and when my conveniency allowed, I endeavoured to compare the Propositions I heard you advance with the Principles of our holy Religion, as contained in the Scriptures of Truth, and agreeably summed up in our excellent *Confession of Faith*.[6]

He found that Hutcheson's propositions and the principles set down in the *Confession of Faith* did not match.

This was a serious matter for Hutcheson, inasmuch as he had taken the oath that all professors in Scottish universities were obliged to take, upon appointment, to teach nothing contrary to the *Confession of Faith.* [7] Heugh knew this: 'I knew you was solemnly engaged before your Admission to your present Office, to assert, maintain and defend all the Truths contained in the said Confession, at least to teach nothing directly opposite to them.' He was also aware that Hutcheson, as a professor of philosophy, was not expected to quote Scripture or employ arguments from revealed religion.[8] But Heugh countered that ideas informed by the light of nature are nothing but 'the faint Remains of that original Knowledge wherewith Man was endowed at his first Creation; that God must therefore be the author of what we know, even in our fallen state; that no one should pretend to knowledge, if that knowledge is inconsistent with Revelation'.[9]

Considered in this light, Hutcheson's teaching was bound to appear heretical. Heugh considered a number of propositions that he had heard Hutcheson utter in the course of his lectures. Proposition I: 'We could have Knowledge of Moral Good and Evil, altho' we know nothing of the Being of God'. The absurdity of this proposition would be evident to any reader of the *Confession,* where he would find Chap. 19, Sects. 1 and 2 that 'God gave to Adam a law, … and this law, after his fall, continued

---

[4] Moncrieff, Alexander, *An Inquiry into the Principle, Rule, and End of Moral Actions* (Edinburgh, 1735), 27.

[5] Glasgow University Archives (thereafter GUA) 26647, 129: Hugh Heugh was awarded a bursary by the faculty of the University on 15 October 1733. He was in the Semi class on logic and metaphysics in 1733–34; he would have taken Hutcheson's class on moral philosophy, for the first time, in 1734–35.

[6] *Shaftsbury's Ghost conjur'd*, 4–5.

[7] GUA 26647, 22, Faculty minutes, 3 November 1730: 'This day the Principal reported that Mr Francis Hutcheson formerly Elected Professor of Philosophy had subscribed the Confession of Faith in presence of the presbytery of Glasgow upon the twenty Ninth of last month'.

[8] *Shaftsbury's Ghost conjur'd*, 5.

[9] *Shaftsbury's Ghost conjur'd*, 6.

to be a perfect rule of righteousness'.[10] And this mistake of Hutcheson's was compounded by Proposition II: 'Tendency to promote the Happiness of others is the Standard of Moral goodness'. Although Hutcheson used the word goodness 'in a certain determined Sense', it was clear to Heugh that the word is 'used very ambiguously, and is capable of various Meanings'. Hutcheson would have found a better understanding of goodness in the law of God, 'for the Law of God is justly called Good, inasmuch as it is the Transcript of his Will, and tends to promote Moral Goodness in these that observe it'.[11] Both propositions were aptly chosen by Hutcheson's critic. The ability to distinguish good and evil, in abstraction from law, human or divine, and the idea of moral goodness as benevolence or a disposition to promote the happiness of others were ideas of central importance in Hutcheson's moral philosophy.[12] There are other propositions cited by Heugh: Proposition III that 'Self-murder is in some Cases lawful' and IV: 'that it is sometimes lawful to make a Lie', where the choice of propositions appears to have been entirely inept. Hutcheson may well have declared, as he is quoted in Proposition V, that 'It is ridiculous to speak of the Sinfulness of Cards and Dice, or any such Diversion wherein Lottery is practised'. Heugh considered determination by lot, as it is mentioned in Scripture to be a solemn matter, not to be treated as a jest.[13]

The propositions that follow are illustrative of more serious differences between Hutcheson and his Reformed or Calvinist critic. Proposition VI: 'It is wrong to say that God acts for his own Glory, or that we ought to have that End always in View'. Heugh cited against Hutcheson the *Westminster Larger Catechism*, question 18: What is God's work of providence? The answer is 'God did decree and create, and does constantly preserve and govern, all things, for his own Glory'.[14] And Hutcheson had declared 'That there is a superiority of Moral Good in the World': Proposition VII. This was not the opinion of the Reformed who thought that in a world inhabited by sinful, fallen men there must be more evil than good in the world.[15] Heugh was sure that there were no grounds for 'such an extensive Charity' as to believe, as Hutcheson did, 'That since Infants are capable of no sin, that all that die before they

---

[10] *The Confession of Faith, Larger and Shorter Catechisms; First agreed upon by the Assembly of Divines at Westminster* (Glasgow, 1732), 42.

[11] *Shaftsbury's Ghost conjur'd*, 9.

[12] Francis Hutcheson, *An Inquiry into the Original of Our Ideas of Beauty and Virtue*, ed. Wolfgang Leidhold, (Indianapolis, Liberty Fund, 2008), 181–2, Treatise II, Section VII, Article V; 'A Synopsis of Metaphysics' Part III, Chapter 4, Article 2, in *Logic, Metaphysics and the Natural Sociability of Mankind,* ed. James Moore and Michael Silverthorne, (Indianapolis, Liberty Fund, 2006), 174.

[13] *Shaftsbury's Ghost conjur'd*, 11–9.

[14] *Shaftsbury's Ghost conjur'd*, 19.

[15] *Shaftsbury's Ghost conjur'd*, 22, quoting *The Confession of Faith* (1732) ch. XVI, Sect. 7, 138: 'Works done by unregenerate Men, although for the Matter of them they may be things which God commands; yet, because they proceed not from an Heart purified by Faith, nor are done in a right Manner according to the Word, nor to a right End, the Glory of god, they are therefore sinful, and cannot please God.'

come to years of discretion will be saved'; for 'altho' Infants are not guilty of any actual Transgression, yet, as they are the Posterity of *Adam*, they are guilty of his first Sin, and are therefore by Nature the Children of Wrath, even as others....'[16]

Here were substantial differences between Hutcheson and his critic. Hutcheson thought that God, who is supremely good, i.e. pre-eminently benevolent, must act for the benefit of others. He did not agree that there must be more evil than good in the world. He thought rather that in a world created by a good God, there must be more good than evil.[17]

And there were deep disagreements between Hutcheson and evangelical Calvinists on the government of the Church. Heugh addressed this matter in the last of his propositions (XI) in which he took it to be Hutcheson's position that 'the Government of the Church belongs to the Civil Magistrate'. Heugh reminded him that this is not the language of the *Confession*, where it is asserted that civil magistrates 'must by no Means assume to themselves the Power of the Keys to the Kingdom of Heaven'[18] Hutcheson would deprive the Church of the power to censure ministers who preached heretical principles. He would have the civil magistrate permit 'unlimited Toleration of all Manner of Doctrines that are not directly inconsistent with public Tranquility'.[19] And he had heard Hutcheson declare that ministers should not be required to subscribe the *Confession of Faith* 'Because (say you) a great many pretty Men, who might be very useful in the church, are by this Means kept out of it being unwilling to bind up their Freedom of Thought in Matters which in their Opinion are not essential to real Goodness'.[20]

It was not only ministers of the Church who should be required to subscribe and adhere to the doctrines enunciated in the *Confession of Faith*, the same requirement should apply to men who are engaged in the instruction of youth. He brought this charge directly to Hutcheson himself and others who had subscribed the *Confession* but taught in a manner directly opposed to it, and 'this being your own Practice, you are no doubt the more ready to judge that others dissemble the same way.'[21] It is a long letter of 42 closely printed pages. Heugh concluded:

> Thus, Sir, I have laid before you a specimen of the Doctrines you teach: And I assure you the eyes of a great many are already upon you, and their Hearts are grieved at the success of your Labours: ...I heartily pity the Youth that are under your charge;... And I earnestly wish that the Lord may preserve them from those dreadful and dangerous Errors;...
> Which is the earnest Desire of
> Your real Well-wisher
> Euzelus Philalethes.[22]

[16] *Shaftsbury's Ghost conjur'd*, 23–4.

[17] Hutcheson, *Inquiry* (2008) II, VII, V, 182: 'And the Deity is call'd good, in a Moral sense, when we apprehend that his whole Providence tends to the universal Happiness of his Creatures; whence we conclude his Benevolence, and Delight in their Happiness'.

[18] *Shaftsbury's Ghost conjur'd*, 35–6, quoting *The Confession of Faith* (1732) chapter 21, Section 3, 54.

[19] *Shaftsbury's Ghost conjur'd*, 36.

[20] *Shaftsbury's Ghost conjur'd*, 37.

[21] *Shaftsbury's Ghost conjur'd*, 38.

[22] *Shaftsbury's Ghost conjur'd*, 41.

11.2 A Vindication of Mr. Hutcheson

Hutcheson's cousin, in Dublin, Will Bruce, was appalled: 'As soon as I saw it I was struck with an apprehension which I have not since been able to get quitt of, that notwithstanding its Impertinence, falsehood and Villainy, It will be able in a great measure to Destroy your Satisfaction and usefulness in your present situation'. Bruce proposed that Hutcheson might do well to return to Ireland to take up a position in the established Church of Ireland. Or, better still, Hutcheson could return to Dublin 'to become the Minister of our new Congregation in Stafford Street', where 'you will be at liberty to teach the principles of Virtue with more freedom than you ever could in your Class and probably with much greater Success in Influencing the Heart'.[23]

A Vindication of Mr. Hutcheson from the Calumnious Aspersions of a Late Pamphlet was published in the same year, 1738, 'by Several of his Scholars'. It was endorsed by 14 men who would vouch for 'the Truth of the Whole of this Account of what Mr. Hutcheson teaches'. They included Hutcheson's devoted students and lifelong friends: Robert and Andrew Foulis and James Moor; Mr. George Rosse, Professor of Humanity in the University of Glasgow; Mr. Gershom Carmichael, Library Keeper; two ministers; two preachers; an elder of the high-church Parish and a merchant and 'we could have mentioned many more'.[24]

In the Preface to the *Vindication* the authors complain of stupidity and lack of charity in Hutcheson's accuser. He could have expressed his misgivings in a private conversation with his professor; or he could have complained to the Dean of the Faculty: 'No such regular Step was taken...'[25] The author was not alone in forming his plan: 'we hear that some in better Stations knew of his Design, perused his paper in Manuscript, encouraged the Design privately, sent the printed paper to their correspondents at a Distance, with their Recommendation'.[26] All of these abettors of the design were animated by religious zeal. The reader is reminded 'that ignorant malicious Zealots have done as much Hurt to the Christian Religion, as any Enemies secret or open'.[27]

A Vindication of Mr. Hutcheson is a curious document. There are two lines of argument and they appear on separate pages. One set of arguments consists of 'Remarks on the Author's Propositions'. These 'Remarks' appear on pages 6, 8, 10, 12, 14, 16 and 18, on the bottom half of page 19 and on page 20. The 'Remarks' form a connected line of argument, linked by the words that appear at the bottom of the page, which anticipate the first word that appears at the top of a page, that follows another page interleaved between them. The interleaved pages are entitled 'Mr. Hutcheson's Propositions'. They are printed on pages 7, 9, 11, 13, 15, 17 and the

---

[23] NLS MS 9252, 128-9, Letter from William Bruce to Francis Hutcheson, 16 December 1738.

[24] *A Vindication of Mr. Hutcheson* (1738) 20.

[25] *A Vindication of Mr. Hutcheson*, 4.

[26] *A Vindication of Mr. Hutcheson*, 3–4.

[27] *A Vindication of Mr. Hutcheson*, 4–5.

top of page 19. These pages restate, in language faithful to Hutcheson's language, what Hutcheson taught. It is an unusual format. It suggests that the 'Remarks' and 'Mr. Hutcheson's Propositions' were composed separately and then assembled after they were printed to juxtapose the 'Remarks' with 'Mr. Hutcheson's Propositions'. Whether Hutcheson himself drafted or was consulted on the propositions attributed to him, it was not Hutcheson but 'Several of his Scholars' who took responsibility for the composition of the *Vindication*. And their concern, as they make clear in the Preface and the conclusion was to provide an accurate representation of what Hutcheson taught.

In response to the accusation I: that Hutcheson taught that one may distinguish between good and evil even if one has no knowledge of God, Hutcheson's defenders agreed. This was indeed his position. For in the absence of an ability to distinguish good and evil, one would have no grounds on which to base a conclusion that God is good or that the Divine Moral Law provides a standard of rightness. II: They acknowledge that Hutcheson's primary notion of moral goodness was not the Divine Moral Law but rather 'Benevolent Affections toward others'.[28] III: They deny that Hutcheson taught that suicide was lawful: 'His grand Aim … was to inspire in his Scholars a noble Contempt of Danger, and a generous Readiness to expose our Lives, were it to the most certain Death, whenever the Cause of our Country, or the Good of Mankind, requir'd it'.[29] IV: He taught that the law of veracity was a sacred precept but 'Almost all Writers on Morals plead, that all these Laws are understood to admit Exceptions, in cases of great Extremity'.[30] V: He did say that it was ridiculous to speak of the sinfulness of cards, dice and lotteries but only in the course of 'his warm Exhortations to his Scholars to abstain from any Diversions which might waste their Time by their being agreeable'.[31]

In reply to the charge that he had declared it 'wrong to say that God acts for his own glory or that we ought to have that end in view' (Proposition VI) we are told that 'Mr. Hutcheson never taught those Words: He showed rather the Ambiguity of the Expression and explained in what Sense God might be said to act for his own Glory'.[32] The sense that Hutcheson attached to those words was that God is good and that 'in the whole of a good God's Works … there must be a Superiority of Good'; he denied 'Bayle's Manichean Objections against the Goodness of God'; he considered that the number of the saved was greater than the number of the damned; and he maintained 'as every Moralist in teaching of the Law of Nature must, that many Actions of Heathens were morally good'.[33]

[28] *A Vindication of Mr. Hutcheson*, 7.

[29] *A Vindication of Mr. Hutcheson*, 9.

[30] *A Vindication of Mr. Hutcheson*, 11.

[31] *A Vindication of Mr. Hutcheson*, 13.

[32] *A Vindication of Mr. Hutcheson*, 13.

[33] *A Vindication of Mr. Hutcheson*, 15. See also Hutcheson, *A System of Moral Philosophy* (1755) I, 9, V.

Finally, on the issue of Church government, his scholars 'remark' that all the charges made against Mr. Hutcheson—that all the powers of the Church derive from the civil magistrate, that heresies in opinion should be expressed without censure, that subscribing a confession should be banished out of the Church were made 'with gross Disingenuity'.[34] But in 'Mr. Hutcheson's Propositions', his position is stated more radically and directly: that 'it belongs to the Magistrate to take care of the religious Notions of the People, to appoint proper Teachers and support them'. He 'also pleaded for universal Toleration by the State, toward all peaceful Subjects of whatever Religion, Let the church censure their Opinions as it pleases: And shewed how this is reconcilable with the Magistrate's Care of Religion.[35]

However one may account for the difference in content and style between the 'Remarks' made by scholars and former scholars and the propositions attributed to Mr. Hutcheson, the latter are generally the more substantive and faithful representations of Hutcheson's teaching. This is particularly the case with respect to his views on Church government. In *A System of Moral Philosophy* (1755; but composed in draft form for distribution to his friends in Ireland and Scotland in the 1730s) and *in A Short Introduction to Moral Philosophy* (1747; but published in Latin in 1742 and 1745) it is asserted with respect to the government of the Church that it is 'the interest of the magistrate and his duty to the state in general to take care that wise and good men be provided and supported to take the leading of such as will be led by some person or other'.[36] He was equally clear that 'the magistrate can have no right to punish any for publishing their sentiments, how false soever he may think them, if they are not harmful to society'.[37] And he defended 'Each one's right of private judgment, especially as to religion' as 'a sense deeply infixed by nature' and therefore 'unalienable'.[38] In the language of the non-subscription controversy in northern Ireland and England, and echoed in Hutcheson's *Inquiry into the Original of our Ideas of Beauty and Virtue*, the right of private judgment was invoked to justify the refusal to subscribe to confessions and creeds such as the *Westminster Confession of Faith*.[39]

It is not surprising that in the 'Remarks' of his scholars and in 'Mr. Hutcheson's propositions' no direct attempt was made to demonstrate that Hutcheson's teaching was consistent with the Westminster Confession of Faith. His understanding of human nature, of morals, of God and the Church were opposed, in virtually all

---

[34] *A Vindication of Mr. Hutcheson,* 18.

[35] *A Vindication of Mr. Hutcheson,* 19.

[36] *A System of Moral Philosophy* (1755) III, 9, I. vol. 2, 311–12 and *A Short Introduction to Moral Philosophy*, ed. Luigi Turco (Indianapolis: Liberty Fund, 2007) III, 8, 1, 267.

[37] *A System of Moral Philosophy* (1755) III, 9, 1, vol. 2, 313 and *A Short Introduction to Moral Philosophy* (2007) III, 8, 1, 267.

[38] *A Short Introduction to Moral Philosophy* (2007) II, 4, 3, 129*;* and *A System of Moral Philosophy* (1755) II, 5, 3, vol. 1, 295.

[39] *An Inquiry into the Original of Our Ideas of Beauty and Virtue* (2008) II, VII, VII, 186. A more extended discussion of Hutcheson's theory of the right of private judgment is provided in Moore 2012.

relevant respects, to the doctrines expounded in the Westminster Confession and in the Larger and Shorter Catechisms. The argument of the *Vindication* was that his teaching was good philosophy; that it represented a defensible understanding of moral goodness, God, divine providence and the Church. But the core argument of *Shaftsbury's Ghost conjur'd* remained: Hutcheson's teaching was directly opposed to the *Westminster Confession of Faith.* It was an argument that was or at least appeared to be unanswerable.

11.3 A Letter to ... Erskine

In the third of his letters to *The Dublin Weekly Journal* on the subject of Bernard Mandeville's *The Fable of the Bees,* Hutcheson considered how one should respond to writing that is 'absolutely Unanswerable'. He recalled Joseph Addison's advice that nonsense cannot be answered or contradicted. One should point out the 'particular excellencies' of the author or utterer of nonsense but in an ironical manner that makes it clear what nonsense it truly is.[40] In the case of *The Fable of the Bees,* it was Mandeville's pretentious claim that he had a deeper understanding of human nature than other philosophers. He attributed this specious claim to his knowledge of anatomy. He had seen, as other philosophers had not, the 'Chief Springs of our Machine' which are yet but 'trifling Films and little Pipes, not such gross things as Nerves, Bone or Skin'.[41] And there were other 'pretended excellencies' in Mandeville's writings: pedantry, astonishing inconsistencies, etc. Hutcheson elaborated upon them. How should one respond to the nonsense offered to the reader by Hugh Heugh?

Sometime subsequent to the publication of *Shaftsbury's Ghost conjur'd* and *A Vindication of Mr. Hutcheson* another open letter was published. It was entitled *A Letter to the Valiant and Undaunted Champion of our Broken Covenants, the Reverend Mr. Ebenezer Erskine, in relation to the present Heresies, Backslidings, Defections, and Lukewarmness of the Times, and his Apostolical Testimonies against them: By a bold young Soldier under his Banner, Euzelus Philalethes, Author of Shaftesbury's Ghost conjur'd.* This open letter is typically attributed, in the libraries that have taken notice of it, to Euzelus Philalethes or Hugh Heugh, as the title page would have one believe. But it would be clear to one who looked beyond the title page that this letter is an ironical celebration of the nonsense contained in *Shaftsbury's Ghost conjur'd.* The anonymous author or authors of this letter celebrate in particular the skill at *conjuring* possessed not only by the 'bold young Soldier'

---

[40] Letter to *The Dublin Weekly Journal,* Saturday, 19 February 1725/6: reprinted in *Hibernicus's Letters,* No. 47, 394–407. The reference is to *The Whig-Examiner,* No. 4, for Thursday, 5 October 1710, reprinted in *The Works of Joseph Addison* (New York, 1837) vol. 3, 265.

[41] Bernard Mandeville, *The Fable of the Bees,* ed. F. B. Kaye (Oxford, 1924) The Preface, iii: quoted by Hutcheson in the third paragraph of his letter to *The Dublin Weekly Journal,* 19 February 1725/6.

in his misrepresentation of the teaching of Hutcheson but also the *conjuring* of the Reverend Ebenezer Erskine and his evangelical colleagues who make nonsense of *The Holy Bible* in their sermons and pamphlets.[42]

In his letter to Erskine, the author boasts of his ability to conjure:

> I find the Vulgar are much offended at my pretending to *conjure*. Conjuring is a diabolical Art, say they: Alas the Blindness of their Hearts! it is one of the prime Graces among the Remnant.[43]
>
> A Conjuror's Business is to transform Things, and to raise Bug-bears and Spectres, that he may have the Glory of laying them again. For instance, I knew well, that neither he, nor any grave Man, ever said, that Lying and Self-murder are lawful. In these Names of mixed Modes, Guilt and Moral Evil are always included: Nay, I know he confuted, at great length, some lax Tenets of Puffendorf and Barbyraque, allowing us to speak against our Opinion in many common Occurrences.[44]
>
> And he showed us too, that scarce any of the noted Instances of Suicide in Antiquity were justifiable ... But to spread abroad, That he taught the Lawfulness of Lying, Murder, and designed Perjury (which last he strongly denied) these were rare Matters of Offence to all who believed my Report; and none but Hereticks can Doubt it...[45]
>
> In good truth, to one who considers him with the Eye of the Flesh, he seems to rejoice, above all Things, in a firm Persuasion of the universal Providence of a Being infinitely wise and good, who loves all his Works, and cannot be conceived as hating any thing he has made. This he constantly inculcates, in the warmest Manner, as a steady Foundation of entire Trust and Confidence in him and chearful Submission to his Will in all Events: That all our Virtues and Advantages of all kinds, are Things that ought never to be ultimately ascribed to ourselves, but to God the Giver of all.[46]
>
> Then, he has such a bewitching Faculty of delivering himself on these Topicks, in a certain simple and striking Manner, that immediately touches the natural Heart, and presents the Imagination with the most beautiful and lovely Forms; that by these Maxims most of all he has exceedingly deluded and debauched the Minds of the unwary youth...[47]

---

[42] The Reverend Ebenezer Erskine (1680–1754) an evangelical minister of great eloquence had taken an active part in 'the Marrow controversy'. In 1732 he preached against restricting the calling of ministers to heritors and elders. In December 1733, he formed the Associate Presbytery with three ministerial colleagues: Alexander Moncrieff, William Wilson and James Fisher. See Lachman 2004 and Vandoodewaard 2011 on the consistency of the theology of 'the Marrow men' with the Westminster Confession of Faith: 62–3 n 13 and 111 ff. on the continuity of the evangelical tradition in the later Churches of the Secession.

[43] *A Letter to ... Erskine* (1738), 10.

[44] The spelling of Barbeyrac's name is irregular in Hutcheson's writings and in the writings examined in this essay. In *A Vindication of Mr. Hutcheson*, 11 and 13, his name was spelled Barbyrac and Barbeyrac. In *A System of Moral Philosophy*, his name was spelled 'Barbeyraque': vol. 1, 264n and vol. 2, 173. And the same spelling, 'Barbeyraque' appears in the MSS of the System: Glasgow University Library (thereafter GUL) MS Gen 110, 168n and 222n. Hutcheson consistently misspelled Pufendorf's name with two 'f's.

[45] *A Letter to ... Erskine*, 13–4.

[46] The understanding of divine providence as infinitely wise and good and the duty to submit cheerfully to God's will were recurrent themes in Hutcheson's natural theology. See 'A Synopsis of Natural Theology' in *Logic, Metaphysics...* (2006) Part III, Chapter 5, 182–5.

[47] *A Letter to ... Erskine*, 15. It was the opinion of virtually all who heard Hutcheson lecture that he was an extraordinarily eloquent and inspiring teacher. William Leechman's Preface to *A System of Moral Philosophy*.

It was not only Heugh himself whom the 'Heugh' of this letter compliments on his conjuring. He compliments Erskine as well for his excellent conjuring: 'You have conjured the present apostatized Kirk of *Scotland* into an Antichristian Synagogue of Christ-dethroners, Covenant-of Works-Upsetters, Off-Christ's-Head-the-Crown-Pluckers, *Erastian*-King-and-Parliament-in-Christ's stead Substituters. In short, without conjuring, there would be no life for the Godly in these Times.'[48]

'Heugh' urges Erskine to 'give this Heretick a second Admonition,' that he revise his manner of teaching entirely; that he ought

> 1st, Never to speak one good word in favour of any Heathen, but to damn them all eternally, Man, woman and child, moral or immoral ... 2dly, to make Tendency to God's Glory the sole Notion of moral Goodness; ...3dly, Instead of vile Pagans, Plato, Aristotle, Cicero, and such other unhallowed Names he recommends, admonish him, as my great Patron does, to study and recommend the *Confession* and *Vincent's Catechism* only. 4thly, To follow the pious Soul-searching Author of the *Fable of the Bees*, ...Or to follow that precious Divine, the great Carnal-reason-Impugner, Mr. Bayle...[49]

The reference to Heugh's patron in the third admonition appears recurrently in the letter. It is a reference to a professor who did no lecturing or writing and appears to have had no scholars except Heugh himself. The same professor is described as the brazier of his Majesty's candlesticks and as someone fascinated by the conjuring word 'Royall'. He may have been the Professor of Ecclesiastical History, William Anderson, who was appointed controversially to the faculty of the University of Glasgow in 1721, by order of King George the First on the advice of the Duke of Montrose, whose sons may have been tutored by Anderson.[50] Hutcheson described Anderson dismissively in a letter to Drennan as 'our Standard and Champion of Orthodoxy' in the University of Glasgow, a distinction he shared with John Loudon, the Professor of Logic and Metaphysics.[51]

The letter to Erskine concludes with a caricature of a sermon that Heugh would like Erskine to read: 'I offer you the Specimen of a Discourse I intend to draw up; by which you'll see my great Abilities in the Interpretation and Application of Scripture, in the manner of true Gospel-preachers, which I have learned from your's and your brethern's Works'. His text is the Book of Judges, Chap. 3, verses 12–22, where he finds 'a clear Prophecy of the present State of the Kirk of Scotland, its Defections, and your Testimony'. It is a wonderfully imaginative reading of Scripture in which the children of Israel are taken to be the Kirk of Scotland; God has raised against them Eglon, a word very like Eglise, which is to say the Kirk; Eglon is king of Moab, which is the broad Hebrew way of pronouncing Mob, which refers to the

---

[48] *A Letter to Erskine*, 11.

[49] *A Letter to Erskine*, 17–8. In the *Inquiry*, 4th edition (1738), Hutcheson observed that 'Plato, Aristotle, Cicero and others of the antients' maintained that 'the natural Principles of Virtue...' are 'implanted in us' (*Inquiry* 2008, p. 244, n. 163).

[50] GUA 26634 fol. 25.

[51] NLS MS 11004 fol. 57–8, Letter from Francis Hutcheson to William Mure 23 November 1743.

Judicature of the Kirk. It is an entirely preposterous outline of a sermon. It occupies the concluding paragraphs of the letter.[52]

Who wrote this letter? Several hands may have been at work on it: among them, perhaps, three of the students who endorsed the *Vindication*: Robert and Andrew Foulis and James Moor, all three were very close to Hutcheson during the remaining years of his short life; very probably, his friend, colleague and relative, Alexander Dunlop, Professor of Greek and a noted wit: Dunlop had a long-standing disdain for Anderson, the Professor of Ecclesiastical History (Scott 1900, 91–2). Another possible contributor to this exercise may have been the Clerk of the General Assembly of the Church of Scotland, William Grant, later Lord Prestongrange, with whom Hutcheson collaborated in the attempt of the Church to confine the calling of prospective ministers to landed gentlemen, i.e. to heritors and elders.[53]

Finally, Hutcheson himself was surely directly involved, not only in much of the writing, but in the conception of the letter as an ironic response to an otherwise unanswerable case. Of course his ideas of human nature, moral goodness, God and the Church were directly opposed to the principles of evangelical Calvinism and the *Westminster Confession of Faith*. The best strategy available to him and to his friends was to celebrate the particular 'excellencies' of the persons who maintained these principles and do so in an ironical manner that made the charges and the arguments they had used against him appear utterly nonsensical.

11.4 The Main Duty of Bishops

One may wonder what Hutcheson would have considered an appropriate subject for ministers of the Church of Scotland to address in their sermons. A very good idea of what Hutcheson thought ministers should preach is expressed in a sermon by the Reverend Mr. Robert Paton, Minister of Renfrew. It was entitled *The Main Duty of Bishops. A Sermon Preached before the Synod of Glasgow and Air [Ayr], October, 1738* (Edinburgh, 1739). Robert Paton was described by another minister, in 1770, as the 'bosom friend' of Francis Hutcheson. Paton is the source of our knowledge that Hutcheson was the author of *Considerations on Patronages* and that he was assisted in the legal part of the pamphlet by William Grant, later lord Prestongrange.[54] Hutcheson, in turn, provided materials for Paton to use in his synod sermon.

[52] *A Letter ... to Erskine*, 19–23.

[53] On Hutcheson's collaboration with William Grant in the composition of *Considerations on Patronages* (1735) see the following note, and Moore 2012.

[54] *Tracts concerning Patronage. By some Eminent Hands*. (Edinburgh, 1770) 'Preface', vii–viii. The editor of this publication, the Reverend Thomas Randall of Stirling, describes how he came to learn that Hutcheson was the author of *Considerations on Patronages* (1735). 'Of this fact the publisher was informed by the late Rev. Mr. Patoun of Renfrew, Dr. Hutcheson's bosom friend, to whom he communicated his manuscript for revisal before publication; who also assured him that in the law part of the pamphlet, the Doctor was assisted by Mr. Grant, late Lord Prestongrange'.

In a letter to Thomas Drennan, without date, but from the context datable sometime after delivery of the sermon in October 1738 but before publication in 1739, Hutcheson wrote:

> The inclos'd you'r not obliged to me for. I was intreated by an old friend who was to preach a Synod Sermon to suggest him some materials, which I undertook … My friend here used a good deal of it in a better method & a diction more suited to this Country, and made an admirable Sermon, but tho' it were printed few would ever dream that had seen the inclosed tho they had read both. And you are only the third person who knows any thing of the Matter.[55]

The use of the term 'bishop' in the title may be confusing. We are told in the first sentence of the sermon that the term is being employed 'in the scriptural sense of that word' to refer to 'the public authorized teachers of the religion of Jesus'. And it was made clear in the opening paragraph that 'By their character and profession, they are the established advocates of virtue and goodness, the fairest, the most amiable, the most honourable and the best thing in the world'.[56] He advised his hearers that they would find this message in the Scriptures: 'These are the pure fountains whence all our instruction must be drawn'. They are likely to be misled if they rely on other sources: 'For what ever stress may be laid upon the authority of primitive fathers, and the venerable decisions of ancient councils; upon the subtle speculations of the schoolmen, or upon established creeds and forms of belief in particular countries, yet as the Bible alone is the religion of Christians, so that alone is the infallible source from whence all our instructions should be derived'.[57] But as one might anticipate, it is not Scripture unaided that should be our guide: 'we should cultivate the sound knowledge of the principles of *natural religion*, and of the *doctrine of manners*; wherein we may derive great assistance from the writings of some eminent Moralists both ancient and modern'.[58] Throughout the sermon ministers are urged to promote 'purity of heart'; 'we should apply ourselves with a peculiar care *to mould and form their hearts* to virtue and goodness; it is the heart which God principally regards'.[59] This was the language of *A System of Moral Philosophy* and of another synod sermon Hutcheson described as 'A Noble one by one of my Scotch Intimates who sees all I do, Mr Leechman.'[60] He also cautioned his fellow ministers that it is 'highly necessary that we ourselves have a *strong relish* of those amiable virtues we teach and recommend to others; it is greatly necessary that our whole souls be animated with a warm zeal to communicate them … if we are cold and indifferent in the cause of virtue, we are but in a poor way of promoting it'.[61]

---

[55] GUL MS Gen 1018, no. 1, Letter to Thomas Drennan, n.d.

[56] *The main Duty of Bishops*, 1.

[57] *The main Duty of Bishops*, 5.

[58] *The main Duty of Bishops*, 8.

[59] William Leechman, *The Temper, Character and Duty of a Minister of the Gospel*, (Glasgow, 1741).

[60] GUL Ms Gen 1018, Letter of Francis Hutcheson to Thomas Drennan, 15 June 1741.

[61] *The main Duty of Bishops*, 16–7. Hutcheson thought it important that one should speak and write in a manner that expressed warmth in the cause of virtue. It is well known that he found such warmth lacking in the moral philosophy of David Hume.

11.5 Epilogue

Hugh Heugh 'died at college in Glasgow, whither he had gone to prosecute his studies for the ministry'; the cause of his death is not recorded (MacGill 1852, 13). Ebenezer Erskine, Alexander Moncrieff, James Fisher and William Wilson seceded with their congregations from the Church of Scotland in 1739.[62] Their numbers were supplemented by other parishes in the decades that followed. A divinity hall was established in Perth by William Wilson in 1737 for the education of future ministers; classes in philosophy were begun in Abernethy in 1742 under Alexander Moncrieff (Whytock 2007, 176–7 and 184–5); the class in moral philosophy at Abernethy was taught from 1749 to 1752 by John Heugh (1731–1810), younger brother of Hugh Heugh (MacGill 1852, 14). The number of seceding Churches continued to increase through the century, culminating in the Disruption of 1843 and the creation of the Free Church of Scotland (Brown and Fry 1993, viii). In the debates that preceded and followed the Disruption, the lives and the writings of the evangelical Calvinists of the 1730s and 1740s were recalled and celebrated in histories and biographies of the 'founders' of the Free Church.

The untimely death of Hugh Heugh appears to have had consequences for the careers of Hutcheson's most vigorous and devoted defenders. In his study of Robert and Andrew Foulis and the Glasgow Press, David Murray reported that 'in the summer of 1738, the brothers, having met with misfortunes by the death of friends with whom they were connected, resolved to make a tour in England and on the Continent' (Murray 1913, 4). Their tour appears to have been connected with a resolution taken by the faculty of the University on 19th January 1738 to obtain from the Scots College in Paris 'a Notorial Copy of the Chartulary of the Cathedral of Glasgow, in which are several papers relating to this University'.[63] They returned to Glasgow to become booksellers and printers. In 1742, Robert Foulis published *The Meditations of the Emperor Marcus Aurelius Antoninus* translated and edited by Francis Hutcheson and James Moor (1742): Hutcheson described it as a work undertaken 'for the sake of a singular worthy soul one Foulis'.[64] Robert and Andrew Foulis would go on to publish 586 books as printers to the University; many were classical texts; 40 of the books were separate editions of the writings of Hutcheson (Ovenden 2004). James Moor was elected Professor of Greek at the University of Glasgow in 1746, mainly due to the influence of Hutcheson (Sher 2004). Robert and Andrew Foulis and James Moor witnessed Hutcheson's last will and testament, dated 30th of June.[65]

[62] *Acts and proceedings of the Associate Presbytery, met at Edinburgh, May 1739, containing their declinature...* Edinburgh, 1739.

[63] GUA 26648, 33.

[64] Letter from Francis Hutcheson to Thomas Drennan, May 31st, 1742: GUL MS Gen 1018, 11.

[65] Public Record Office of Northern Ireland (PRONI): D/971/34/D/1.

There were no doubt other moments, in the eighteenth century, in Scotland, when philosophers, ministers and men of letters whom we have come to identify as enlightened found themselves confronted by evangelical Calvinists. In this essay, we have seen how Francis Hutcheson, his students and former students, colleagues and friends responded to such a challenge. Their side of the debate was conducted in a *Vindication* endorsed by 14 of Hutcheson's students and former students; in an open letter to the leader of the evangelical Calvinists which took the form of an ironic celebration of the excellence of conjuring; and in a synod sermon composed by Hutcheson's 'bosom friend' using materials provided by Hutcheson. In all of these literary productions, responsibility for authorship was taken not by Hutcheson but by others. This is consistent with the style in which Hutcheson, a prudent and cautious man, by his own account as well as by the testimony of others, found it appropriate to conduct his controversies in the zealous and repressive environment of Glasgow in the 1730s and 1740s.

References

Brown, Stewart J., and Michael Fry. 1993. *Scotland in the age of the disruption*. Edinburgh: Edinburgh University Press.

Hutcheson, Francis and James Moor. 1742. *The Meditations of the Emperor Marcus Aurelius Antoninus. Translated by Francis Hutcheson and James Moor*. 2008. Edited and with an Introduction by James Moore and Michael Silverthorne. Indianapolis: Liberty Fund.

Kidd, Colin. 2000. Scotland's invisible enlightenment: subscription and heterodoxy in the eighteenth-century Kirk. *Records of the Scottish Church History Society* 30: 28–59.

Lachman, David C. 2004. Erskine, Ebenezer (1680–1754). In *Oxford dictionary of national biography*. Oxford: Oxford University Press. http://www.oxforddnb.com/view/article/8853. Accessed 14 May 2011.

MacGill, Hamilton. 1852. *The life of Hugh Heugh*, D.D. Edinburgh: A. Fullarton and Co. (The subject of this biography, Hugh Heugh D. D. (1782–1846) was the son of John Heugh (1731–1846), younger brother of Hugh Heugh (d. 1738)).

Moore, James. 2012. Presbyterianism and the right of private judgment: church government in Ireland and Scotland in the age of Francis Hutcheson. In *Philosophy and religion in enlightenment Britain*, ed. Ruth Evelyn Savage, 141–68. Oxford: Oxford University Press.

Murray, David. 1913. *Robert and Andrew Foulis and the Glasgow Press with some account of the Glasgow Academy of the Fine Arts*. Glasgow: James Maclehose.

Ovenden, Richard. 2004. Foulis, Robert (1707–1776). In *Oxford dictionary of national biography*. Oxford: Oxford University Press. http://www.oxforddnb.com/view/article/9991. Accessed 14 May 2011.

Scott, W.R. 1900. *Francis Hutcheson: his life, teaching and position in the history of philosophy*. Cambridge: Cambridge University Press.

Sefton, Henry. 1983. Neu-lights and Preachers Legall: some observations on the beginnings of Moderatism in the Church of Scotland. In *Church, politics and society in Scotland 1408–1929*, ed. Norman McDougall, 187–96. Edinburgh: John Donald Publishers Ltd.

Sher, Richard B. 1985. *Church and university in the Scottish enlightenment: The moderate literati of Edinburgh*. Edinburgh: Edinburgh University Press.

Sher, Richard B. 2004. James Moor (b. 1712–1779). In *Oxford dictionary of national biography*. Oxford: Oxford University Press. http://www.oxforddnb.com/view/article/19090. Accessed 14 May 2011.

Vandoodewaard, William. 2011. *The marrow controversy and seceder tradition: Marrow theology in the associate presbytery and associate synod secession churches of Scotland (1733–1799).* Grand Rapids: Reformation Heritage Books.

Whytock, Jack. 2007. *'An educated clergy': Scottish theological education and training in the Kirk and secession, 1560–1850*. Milton Keynes: Paternoster.

Chapter 12
Questioning Church Doctrine in Private Correspondence in the Eighteenth Century: Jean Bouhier's Doubts Concerning the Soul

Ann Thomson

The study of the circulation of irreligious ideas in France in the eighteenth century, particularly in the form of what are called 'clandestine manuscripts' but should more properly be termed clandestine literature, has greatly developed in recent years.[1] These irreligious texts, which indeed circulated often in the form of manuscripts, some of them copied from scarce printed works, were recopied and altered as they passed from person to person. They constituted a fund of heterodox ideas and erudition which was used throughout the eighteenth century. Their study has helped to throw light on heterodox speculation and its circulation among the intellectual elite, although many problems remain concerning in particular the authorship and dates of composition of many manuscripts. The works themselves are very varied, draw on a variety of ancient and modern sources, and are far from representing a single philosophical or religious position; nevertheless their study has helped to reinforce belief in certain quarters in a coherent 'radical enlightenment' composed of a group of like-minded thinkers organized into a network.[2] In this context, scholars have emphasized the continuity of an irreligious tradition drawing on a certain number of philosophical texts, which helped to undermine not only religious doctrine but also political authority. Research has also, however, shown that theological works were frequently mined for the irreligious arguments to which their authors were replying and that theological debates could fuel this underground literature.

[1] See in particular Benitez 1994, Thomson 1981.

[2] Jacob 1981; Israel 2000.

A. Thomson (✉)
Department of History and Civilisation, European University Institute,
Via Boccaccio 121, 50133 Firenze, Italy
e-mail: ann.thomson@wanadoo.fr

A. Dunan-Page and C. Prunier (eds.), *Debating the Faith: Religion and Letter Writing in Great Britain, 1550-1800*, International Archives of the History of Ideas 209,
DOI 10.1007/978-94-007-5216-0_12, © Springer Science+Business Media Dordrecht 2013

The present chapter will go further in this direction by looking at an example of the preoccupations of certain sincere Christians whose worries about some Church doctrines led them to voice heterodox speculation in their private correspondence. We shall see on the one hand the influence of the theological debates that could be conducted more openly across the Channel in the British Isles, and on the other the closeness of these doubts to those expressed in some clandestine literature.

The particular correspondence studied here is that of Jean Bouhier (1673–1746), 'Président à mortier' in the Dijon Parlement, member of a leading Dijon family which included other Parlementaires, Church dignitaries and so on, and a prominent and respected member of the republic of letters. He was a jurist but also an antiquarian and scholar in several fields of erudition. He was elected to the Académie française in 1727, and despite the condition that members should reside in Paris he remained all his life in Dijon without visiting Paris again. He conducted an extensive correspondence with a large number of scholars, particularly after ill health forced him to sell his position in the Dijon Parlement in 1726. The inventory of this correspondence was drawn up over 30 years ago by Françoise Weil,[3] but although some of the letters were published by Henri Duranton, particularly the exchanges concerning literary news and discussions of recent publications,[4] a large number remain unpublished in several huge volumes at the Bibliothèque nationale de France. They contain essentially Bouhier's passive correspondence, but there are a certain number of drafts of his replies, which enable us to understand his private opinions.

As can be seen from the brief presentation of his life given above, Jean Bouhier was a pillar of the establishment, both social and intellectual, as of the Catholic Church. After having defended the Dijon Parlement in the 1720s against the encroachment of its power by the royal government, he seems to have altered his stance later; in the agitation in the early 1730s concerning the Papal Bull *Unigenitus* condemning the Jansenists, he finally opposed the Jansenist party and disagreed with the Paris Parlement's resistance to royal authority.[5] As a result, his nineteenth-century biographers and Michaud's *Biographie universelle* emphasize his unquestioning faith. However, while Bouhier himself insisted on the danger of openly questioning religion and disapproved of works which would undermine institutions and the social order, Françoise Weil points out that in private he did permit doubts concerning certain Church doctrines. He also seems to have believed in the legitimacy of examining dogma and the basis of belief; as he wrote to the Protestant pastor Iselin: 'il est permis et même convenable à tout chrétien qui en est capable d'examiner dans leurs sources les dogmes de la foi'.[6] F. Weil refers specifically to the question of the nature of the soul and she links discussion of this question to the clandestine philosophical manuscripts that circulated in France during these years, which were

[3] Weil 1975.

[4] *Correspondance littéraire du Président Bouhier* (Saint-Etienne: Université de Saint-Etienne, 1976–).

[5] Weil 1975, 1: XXXVIII.

[6] Quoted by Weil 1975, 1: XVLIII.

one of the main vectors for the diffusion of irreligious, or at least anti-catholic, arguments.[7] While her claim that Bouhier had seen some of these works is perfectly plausible the present chapter, based on what he wrote in his private correspondence, will investigate a different source of speculation casting doubts on Church doctrine concerning the nature of the soul—a source which itself also fuelled the arguments of clandestine irreligious works, as I have shown elsewhere.[8]

While the speculation on the soul was part of an irreligious 'libertin' tradition derived from Epicureanism and Stoicism and still drawing on the classical texts, it had also strong and ancient roots within Christianity, often influenced by the same sources. By the late seventeenth century certain heterodox views of the soul were particularly influential inside various Protestant Churches. As it is impossible, in the context of this chapter, to go in detail into this question and its origins, the present discussion will be limited to a brief mention of the debate in England in the late seventeenth century, during the unsettled years following the Glorious Revolution and the lapsing of the licensing Act in 1695, whose effect was to permit uncensored publication in England. This happened at the same time as, and doubtless facilitated, the Tory High Church offensive against toleration on the one hand, and the 'True Whig' attack on priestcraft on the other.[9] The Socinians also launched a campaign for a return to purer Christian doctrines within the Anglican Church.[10] One of the doctrines criticized by Socinians and others was the Catholic teaching concerning purgatory linked to belief in the necessary immateriality of the human soul as a precondition for its immortality. This doctrine, reaffirmed in the 1513 Lateran Council, was considered to be a pagan Platonic importation. The effect of Cartesian dualism and the works of those usually called the 'Cambridge Platonists', in particular Ralph Cudworth and Henry More, was to reinforce its importance. The demonstration of the soul's immateriality became the foundation of the proof of its immortality, based on the distinction between the two substances. In reply to the arguments of Hobbes and the Epicureans, theologians and philosophers insisted on the necessary immateriality of thinking substance and the impossibility, even for God, to accord matter the capacity for thought.

Such insistence led in turn to a reaction from certain 'Christian mortalists', who revived a tradition which had received some support from Luther. They claimed that on death the whole human being died (or slept, according some believers) until Judgement Day when the whole person was resurrected and entered into immortality. Locke's speculation on thinking matter needs also to be understood in this context. The main defenders of such a position in the 1690s and early years of the eighteenth century were a pious old gentleman called Henry Layton and Dr William Coward, whose writings in defense of a wholly material conception of humans also drew on medical evidence concerning the human brain and living matter. According to them

[7] Weil 1975, 2: XLVIII.

[8] Thomson 2007.

[9] Rose 1999, 182 ff.; Goldie 1993.

[10] Trowell 1996.

their position conformed to both Biblical teaching and the views of the Church Fathers, in particular Irenaeus or Tertullian. The high-profile polemic they launched, which is also the context of John Toland's *Letters to Serena* (1704), raged during the first decade of the eighteenth century, with exchanges of pamphlets and denunciations from the pulpit. The High Church party and the Tories attempted to use such works to attack the Latitudinarian bishops and reverse the laws on toleration, and they obtained the condemnation of Coward's works by the House of Commons.[11]

The whole debate was followed with great interest in French language periodicals, not only those published by the Huguenots in Holland—who were naturally particularly interested in Protestant theological debate—but even such mainstream journals as the *Journal des Sçavans*. Particular attention was accorded in France to the final phase of this polemic, around the works of Henry Dodwell, a highly respected Biblical scholar and theologian who became a non-juror at the accession of William and Mary and finally left the Church in protest against the appointment of bishops who had sworn allegiance to William and Mary in replacement of the non-jurors. Having lost his Oxford post, he settled nearby and became the centre of a network of Jacobites, but also continued to be consulted on questions of erudition by all sorts of scholars. He published in 1706 *An Epistolary Discourse proving from the Scriptures and the First Fathers that the Soul is a principle naturally mortal but immortalized actually by the pleasure of God to punishment, or to reward, by its union with the divine baptismal spirit, wherein is proved, that none have the power of giving this divine immortalising spirit since the apostles but only the bishops.* As the title makes clear, he was not denying the immateriality of the soul but its natural immortality, in order to argue that the illegitimate Church of England bishops who had replaced the non-jurors did not have the power to immortalize souls by baptism. The work was thus part of his campaign against the Church hierarchy.[12] Despite this, his views were not surprisingly compared to those of Layton and Coward and he was accused of irreligion and freethinking. A refutation by Samuel Clarke led to a famous exchange of pamphlets between Clarke and the freethinker Anthony Collins, a notable critic of 'Priestcraft', who seized the opportunity to deny a separate immaterial and immortal soul from a very different standpoint.[13]

Although these questions seem abstruse today, they were at the heart of Christian doctrine and the question of the soul was seen to be essential for demonstrations of the existence of God and belief in Providence. Hence the interest they aroused, and the large number of refutations, pamphlets, sermons, or articles devoted to the question. Both Daniel Defoe and Jonathan Swift refer to it in several of their writings in this period.[14] As already indicated, all of this was followed with great interest in France and many French-language periodicals carried reviews of

[11] For details see Thomson 2008b, 97–134.

[12] See Leighton 2002.

[13] See Vailati 1997, 54–62.

[14] Daniel Defoe, *The Consolidator: Or Memoirs of Sundry Transactions from the World in the Moon* (1705) 33, 92–9; Jonathan Swift, *A Tale of a Tub With other early Works 1696-1707* ed. Herbert Davis (Oxford: B. Blackwell, 1957) 71, 72, 95.

Dodwell's works. Both the Jesuit *Journal de Trévoux* (which gleefully pointed out that Dodwell's only solution was to return to the Catholic Church)[15] and the Huguenot periodicals devoted many pages to his books, thus enabling his ideas to be well known to the French public. As a result, references to Dodwell's arguments appear in at least one irreligious clandestine work, as do quotes from Swift's reaction to them.[16] But what is of particular interest in the context of the present chapter is the way in which, in France as in England, this speculation was not confined to enemies of religion. On the contrary, in the early decades of the eighteenth century it nourished the doubts of sincere Christians anxious to understand God's will and worried about aspects of Catholic teaching. They seem to have been dissatisfied with some of the Church's answers to questions concerning not only salvation but also the nature of the deity and of his creation. In view of censorship, in France such doubts could not be expressed openly as in England but were confined to private discussion or correspondence.

This seems to be the case for Jean Bouhier (1678–1742), who discussed the nature of the soul in exchanges of letters with several Protestants; this detail is in itself interesting, and shows that the intellectual networks of the republic of letters transcended confessional differences. He began around 1733 a correspondence with the Leibnizian scientist Louis Bourguet, a Huguenot from Nîmes who had settled in Switzerland in Neuchâtel. Bourguet was a very respected figure who exchanged letters with many leading scientists and seems to have given Buffon the idea for his 'moules intérieures', a theory developed in *Lettres philosophiques sur la formation des sels et des crystaux et sur la génération & le méchanisme organique, des plantes et des animaux* published in 1729. Although this work discussed mainly questions of mineralogy, which was his principal interest and the subject of other works by him, several of the letters in this volume concerned the important question of reproduction, which was the object of much speculation and experiment in the early eighteenth century. Here Bourguet defended Leibniz's theories on preformation and countered certain ideas which he saw to be dangerously irreligious and smacking of Spinozism.[17] He elaborated on these ideas in the letters he wrote to Bouhier in 1736 and 1737, defending Leibniz's pre-established harmony and view of the soul and criticizing in particular Spinoza but also Locke for encouraging irreligious ideas. According to him, to accept a material soul is 'du Spinozisme tout pur'.

In November 1737 Bourguet attacked violently those who like Locke envisage the idea that matter might think:

> J'avoue, Monsieur, qu'en admettant que Dieu peut donner à la Matiere la faculté de penser, les Materialistes n'y gagneroient rien, puis qu'il dependroit dans ce cas là de la volonté de Dieu de conserver l'ame quoique materielle. Mais ces Mrs. ne sont Deistes que de nom

[15] *Journal de Trévoux*, July 1706, 1260–1.

[16] See Thomson 2013.

[17] Louis Bourguet, *Lettres philosophiques sur la formation des sels et des crystaux et sur la génération & le méchanisme organique, des plantes et des animaux ; à l'occasion de la Pierre Belemnite et de la Pierre Lenticulaire, avec un memoire sur la Théorie de la Terre* (Amsterdam: François L'Honoré, 1729).

dans le fond, ce sont de vrais Spinosistes, de vrais Pantheistes. Ils ne parlent le langage des Orthodoxes (j'appelle ainsi ceux qui ont des idées saines de Dieu) que pour leurrer le monde. Ils aiment mieux penser que le Moy se perd apres la destruction du corps, que de l'envisager comme subsistant. Or si comme Spinosa l'imaginoit tout est matiere, Dieu & l'univers sont une même chose.[18]

Bourguet is here expressing a widely held view concerning the need to distinguish immaterial thinking substance from matter, as without this distinction the whole edifice would collapse. Bouhier however did not agree with Bourguet and in his reply, written on 4 December 1737, his criticism of pre-established harmony is followed by this remark:

Pour le matérialisme de l'ame, il est bien vrai, que quelques libertins en peuvent abuser. Mais si dans le fond, il n'a rien de contraire à la Religion, l'abus qu'on en peut faire, n'est pas une raison pour le rejetter. On sçait, que ça été le sentiment de plusieurs anciens Peres. On peut donc l'admettre, pourvû qu'il ne renferme rien de contradictoire.[19]

The view that Bouhier is here expressing privately was—as is evident from Bourguet's statements, from the reaction to Locke's hypothesis and from the large number of theological works directed against free-thinking tendencies—considered highly unorthodox and positively dangerous for belief and for society. Nevertheless, this was far from a passing remark.

Bouhier set out in more detail the arguments and justifications for his views in his exchange of letters with another correspondent in Neuchâtel called Caspar Cuenz. This exchange is particularly instructive as it is more in the nature of an investigation than a statement of fixed positions. Caspar Cuenz was a close friend of Bourguet but was hardly known in his own time or since, despite the fact that he conducted a voluminous correspondence with several leading scientists and members of the republic of letters. Cuenz (1676–1752), was 'conseiller d'état' in the Swiss town of Saint-Gall and also member of the Marseille Académie des Sciences et Belles-Lettres. After representing his town abroad, notably in London, he retired to Neuchâtel in 1738. It was here that he could finally indulge a passion for metaphysics that he had acquired when he started to read Descartes and other philosophers in Paris from 1722 onwards.[20] In Neuchâtel he met Louis Bourguet, whose attempts to

[18] Bibliothèque nationale de France (thereafter BNF) mss. f. fr. 24409, fol. 251, v. I concede, sir, that by admitting that God can give matter the faculty of thinking, the materialists would gain nothing, for in this case it would depend on God's will to conserve the soul even though it is material. But these gentlemen are only deists in name, at heart they are true Spinozists, true Pantheists. They speak the language of the orthodox (by which I mean those who have healthy ideas of God) only in order to deceive people. They prefer to think that the self is lost after the body's destruction rather than envisage it subsisting. But if, as Spinoza imagined, everything is matter, then God and the universe are the same thing.

[19] BNF mss. f. fr. 24409, fol. 248 v. As for the soul's materialism, it is true that a few libertines can take advantage of it. But if at root it is in no way contrary to religion, this misuse is not a reason to reject it. We know that it was the opinion of several former Fathers. We can thus admit it provided it does not involve a contradiction.

[20] See his letter to Bouhier, BNF mss. f. fr. 24410, fol. 243.

convert him to Leibniz's philosophy led Cuenz to develop his own system. Cuenz worked out over several years a theory of spiritual beings based on Locke's philosophy, which he expounded in four rambling and disorganized volumes published in 1742, probably at his own expense, under the title of *Essai d'un sisteme nouveau concernant la nature des etres spirituels*. This book attempted to provide a material explanation for human beings which dispensed with the need for an immaterial soul, and to defend the idea of an extended God. According to him immaterial substances are 'pure beings of reason' as according to the Biblical account God breathed life into material beings. In his view: 'Il n'y a pas lieu de douter que Dieu n'ait prédisposé cette machine humaine, pour la rendre, au moyen de ce souffle de vie, capable de la pensée, du sentiment et de tout ce qui en dépend'.[21]

Cuenz believed that his system was the only one capable of countering the irreligious arguments of freethinkers. In order to promote his ideas he consulted scholars and scientists all over Europe and submitted his manuscript to the President of the Royal Society. Hans Sloane commissioned a report, which dismissed the work as that of a worthless eccentric. This view seems to have been shared by several scholars while others, including Charles Bonnet or Jean Henri Samuel Formey, secretary of the Berlin Academy, conducted a long correspondence with him on these questions.

I have studied elsewhere the curious case of Cuenz, which throws much light on the workings of the republic of letters and science, and on correspondence networks.[22] The present chapter will concentrate, however briefly, on his exchange with Bouhier and the long letters Cuenz wrote in somewhat defective French between 1738 and 1745. These letters have been known for some time and their interest was pointed out by Ira O. Wade, whose approach to them insisted on the parallels with Voltaire's thought (Wade 1975). In my opinion such an approach misses the vital theological component of the debate.

Bouhier and Cuenz seem to have met during a visit made by the Swiss official to France in 1729, when he was working on a project connected with the Bourgogne Canal.[23] As Cuenz had apparently understood via Bourguet that Bouhier shared his ideas, he resumed contact with the Dijon magistrate in 1738 in order to ask for the opinion of this leading light in the republic of letters concerning his system. He informed Bouhier that Bourguet had not succeeded in converting him to Leibnizianism: 'il y a une amitié intime entre M. bourguet et moy; mais il n'y a pas apparence, que nous nous convertissions jamais l'un l'autre concernant la Metaphisique'.[24] Bouhier replied that he had abandoned metaphysical speculation

---

[21] *Essai d'un sisteme nouveau*, I, 14. It is beyond doubt that God has predisposed this human machine in order, by means of this breath of life, to make it capable of thought, feeling and everything that depends on them.

[22] Thomson 2008a. See also Yolton 1991, 76–84.

[23] See BNF mss. f. fr. 24410, fol. 242.

[24] BNF mss. f. fr. 24410, fol. 243, 288 v. M. Bourguet and I are intimate friends; but there is no sign that we will ever succeed in converting each other on the subject of metaphysics.

due to its uncertainty and the disagreements among philosophers, and thus that he could consider impartially Cuenz's ideas, and this led to an interesting exchange of letters as the Frenchman gave advice to the aspiring metaphysician.

Cuenz expounded his system based on two principles:

> qu'un Etre non étendu et purement spirituel capable du sentiment, de Puissance et de toutes ses modifications est une contradiction palpable, et qu'il suffit de distinguer les Etres qui existent en visibles et palpables, et en invisibles et impalpables à nos sens Grossiers, hypothèse tres aisée à concilier avec la revelation et avec les grandes fins qu'eux nous proposent, et plus propre que toute autre à fermer la bouche aux Pyrrhoniens et aux Esprits forts.[25]

His ideas were derived from his understanding of Locke and were particularly influenced by English writings, due no doubt to his stay in England. While he was there, his friend the deist journalist and member of the Royal Society Thémiseul de Saint-Hyacinthe seems to have provided him with many works, including those of John Toland, and probably translated them into French for him.[26] Many of the self-taught Cuenz's theories were deeply heretical, despite his desire to defend religion and reduce freethinkers to silence. He denied not only the existence of immaterial substance but also the creation, as did the Epicureans:

> Je ne crois pas que ce soit une heresie ou une absurdité de supposer que la matiere est ab æterno. La revelation ne dit nulle part que Dieu a tiré la matiere du néant, ou pour me servir d'une Expression moins equivoque, que Dieu par un simple acte de sa volonté a donné l'Etre à la Matiere.[27]

He sent Bouhier his system, in reply to which Bouhier wrote a long commentary in 1738, with a certain number of questions and some advice for further reading, including Beausobre's *Histoire du manichéisme*. While Bouhier rejected out of hand Cuenz's claim that God was material, as it led to insuperable difficulties, he had no difficulty with a material soul:

> Pour tous les autres Etres, je n'y vois point le même inconvénient, & c'est ce qui m'a fait avoir souvent des idées pareilles aux vostres, quoique je ne me sois jamais appliqué à les réduire en systeme, n'en ayant ni la capacité, ni le loisir. mais je suis persuadé de vostre grand principe que, Dieu a sans doute préféré les voyes les plus simples aux composées, & qu'il a eté aussi aisé à la toute puissance divine de faire des machines pensantes qu'il l'est aux hommes d'en faire d'une autre espèce, & de porter même l'industrie jusques à des choses, qui paroitroient incroyables, si nous ne les avions vûes. Je mets de ce nombre les Tableaux mouvans, que vous connoissez sans doute, & ce Flûteur artificiel, inventé depuis

[25] BNF mss. f. fr. 24410, fol. 245, 9 April 1738. that a non extended and purely spiritual being capable of feeling, power and all its modifications is a palpable contradiction, and that it is enough to distinguish existing beings into visible and palpable, and invisible and impalpable to our crude senses, a hypothesis very easy to reconcile with Revelation and with the great aims they propose, and more suitable than any other of closing the mouths of pyrrhonists and freethinkers.

[26] Carayol 1984.

[27] BNF mss. f. fr. 24410, fol. 290. I do not think it is heretical or absurd to suppose that matter is ab æterno. Revelation nowhere says that God drew matter from nothingness, or to use a less equivocal expression, that he gave being to matter by a simple act of will.

peu par un jeune homme, qui joue de la flûte Allemande jusques à 14 airs différrends, avec toute la precision, & la justesse possible. Refusera-t'on à dieu l'habileté de faire des machines infiniment supérieurs?

A la vérité, il est difficile de concevoir, comment une chose, qui nous paroit incorporelle, telle que la pensée, peut être opérée par des organes purement corporels. Cependant la plupart de nos philosophes, qui accordent aux bêtes la faculté de penser, en même tems qu'ils croyent leur ame matérielle, nous ont en quelque maniere frayé le chemin pour en dire autant des hommes.

La religion bien entendue ne doit point mettre d'obstacle à cette opinion. Car on ne sçauroit refuser à dieu le pouvoir de resusciter la matiére & de la rendre immortelle, & par consequent capable de récompenses, ou de punitions éternelles. on scait même que les anciens Peres croyoient la materialité de l'ame sans croire blesser la religion. Ce sera pourtant le point le plus délicat à tomber. Mais cette difficulté n'est point insurmontable pour un aussi bon esprit que le votre (fol. 264 r–v).[28]

Two years later he explained his ideas even more clearly:

Il seroit à souhaiter, que ce qui regarde la nature de nôtre ame fût aussi aisé à eclaircir. comme j'y ai toujours trouvé de grandes difficultez, il m'a paru que le mieux etoit de suspendre notre jugement sur ce point, & de nous en tenir par provision aux idées communes, dans lesquelles nous avons eté elevez.

Je suis pourtant persuadé, autant que mes foibles lumiéres peuvent m'eclairer, que le système de la matérialité de l'ame, qui semble avoir eté celui des premiers Peres de l'Eglise n'a rien de contraire à notre Religion, & qu'il peut fort bien etre soutenu en bonne Philosophie. Vous sçavez que je me suis déjà quelques fois expliqué sur ce dernier point. Et à l'egard de l'autre, il me paroit qu'en faisant voir, que dieu a pû rendre notre ame immortelle, quoique matérielle, & par consequent capable de récompenses ou des peines eternelles, il n'en resulte aucun inconvénient dangereux pour les dogmes du christianisme.[29]

---

[28] I do not see the same drawback for all the other beings, and that is what often gave me ideas similar to yours, although I have never tried to reduce them to a system, lacking as I do both the capacity and the leisure. But I am convinced of your great principle that God no doubt preferred the most simple means to complicated ones, and that it was just as easy for divine omnipotence to make thinking machines as it is for men to make machines of a different type, and even to extend their efforts to things which would seem incredible if we had not seen them. I include moving pictures, which you doubtless know, the artificial flute player, invented recently by a young man, who plays on the German flute as many as 14 different tunes, with all the precision and accuracy possible. Can we refuse God the skill to make infinitely superior machines? § In truth, it is difficult to conceive how something which seems incorporeal to us, like thought, can be operated by purely corporeal organs. Yet most of our philosophers, who admit the faculty of thought in animals while believing their soul material, have in some sort prepared the way for us to say the same of men. § Religion properly understood should not put obstacles in the path of this opinion. For we cannot refuse God the power to resuscitate matter and make it immortal, and as a result capable of eternal rewards or punishments. We even know that the Fathers believed in the materiality of the soul without believing they were harming religion. This would however be the most delicate point to win. But this difficulty can be overcome by a mind as good as yours.

[29] été 1740, fol. 304. I would like it to be as easy to throw light on the nature of our soul. As I have always found that this question involved great difficulty, I believed that the best thing was to suspend our judgement on this point and to stick for the moment to the common ideas with which we were brought up. § I am however convinced, as far as my weak understanding can enlighten me, that the system of the soul's materiality, which seems to have been that of the first Church Fathers, is in no way contrary to religion, and that it can be very well defended in good philosophy.

Bouhier is here reacting to philosophico-theological discussions which had exercised minds over a long period. On the one hand, there is the question of God's omnipotence: as we have seen above, many theologians and others insisted that the very nature of matter was such that even God could not make a particular organization of matter think. The first Boyle Lecturer Richard Bentley proclaimed in 1692: 'Omnipotence itself cannot create cogitative body. And 'tis not any imperfection in the power of God, but an incapacity in the subject. The ideas of matter and thought are absolutely incompatible.'[30] Any denial of this opened the door to atheism and Spinozism. Such a position, which Locke contested at length in his exchange of pamphlets with Edward Stillingfleet, Bishop of Worcester, in 1696–1697, shocked profoundly many Christians as it seemed to limit God's power. As we have seen they claimed, following the Bible, that God would raise the dead at the Last Judgement when they would enter on immortality. Denial of a separate immaterial soul did not therefore mean a denial of immortality or of divine punishment and recompense in the afterlife. This position seems to be the one defended here by Bouhier, despite its link to Protestant thinking.

It is interesting to observe that he cites in defense of this view the complex machines made by eighteenth-century engineers which could imitate human actions and which so fascinated contemporaries. He is not however comparing humans to such machines, but is claiming that if humans can make such amazing machines out of matter, then we cannot deny God the ability to make infinitely more perfect material beings, such as humans. This is not therefore an extension of the Cartesian animal-machine hypothesis to humans, which has often been considered to be the basis for materialistic interpretations of humans such as La Mettrie's *Homme machine* (1747).[31] On the contrary, it is linked to the claim that matter itself can account for all human faculties and that a material soul can exist and thus to a denial of two substance dualism.

In this connection, the question of animals inevitably came up, as the issue of animal souls was a particularly tricky one for theologians and one which was raised on both sides of the dispute.[32] Pierre Bayle rehearsed at length the various arguments about animal soul in several articles in his *Dictionary*, in particular 'Rorarius', to which Cuenz refers. According to Bayle, either one accepted the 'absurd' Cartesian solution that animals were unfeeling and unthinking purely material beings or 'machines' (a solution which involved many difficulties and

---

You know that I have already explained my position on this last point several times. And as for the other, it seems to me that by showing that God could have made our soul immortal despite its being material, and as a result capable of eternal rewards or punishments, no dangerous consequence follows for Christian dogma.

[30] Richard Bentley. *Eight Sermons Preached at the Honorable Robert Boyle's Lecture…*, 5th ed. (Cambridge, 1724) 75.

[31] On this subject, see Thomson 1988.

[32] See Thomson 2010.

was difficult to defend, as Bayle admits), or one had to accord them some type of material sensitive soul and take the dangerous step of abandoning strict two substance dualism.[33] In view of the faculties demonstrated by many animals, it was difficult not to accord them some type of rudimentary thought. Once that step had been taken, as Bouhier points out, it was easy to move on to accept the idea that a superior material system could account for the human soul. This does appear to be Locke's position; for him the difference between human and animal souls, like the difference between their faculties, seems to be one of degree and not kind.[34] The question came down to that of thinking matter and the existence of a necessarily immaterial thinking substance.

Bouhier is perfectly aware that the opinions he expresses are not generally acceptable, despite his belief that they are not counter to religious teaching, and while agreeing to let the Swiss metaphysician publish at the end of his work his own letter to Bourguet, he insisted that his name should not appear (fol. 304).

After reading Cuenz's system in July 1740, Bouhier expresses disagreement concerning the specific view of the soul it defends, which appears to him to be contradictory. According to Bouhier:

> Il me semble que vous auriez mis votre système à l'abri de toutes ces contradictions, en soutenant simplement, que l'ame est une modification de la matiere, fait d'une maniére inperceptible par l'Etre suprème, qui lui a donné la faculté de penser, d'apercevoir, &c, d'où il s'ensuit, qu'elle est mortelle de sa nature; mais renduë immortelle par la volonté divine (fol. 304).[35]

He informs his correspondent that this was the opinion of 'one of the most learned Englishmen of recent times, I mean the famous Henry Dodwell', whom he advises him to read. He doubtless knew of Dodwell's fame as a classical and Biblical scholar and thus was particularly open to his theological ideas, although according to Françoise Weil, Bouhier probably did not know foreign languages and certainly not English.[36] Here we see clearly the role played by the various journalists who had provided long reviews of and extracts from Dodwell's works. In fact, Bouhier directs Cuenz to the extract from Dodwell's book that appeared more than 30 years earlier in Henri Basnage de Beauval's *Histoire des Ouvrages des Scavans*, for which he provides the complete reference and page number.[37] He adds another reference,

[33] Pierre Bayle, article 'Rorarius', rem. C, D, in *Dictionnaire Historique et critique*, 5th ed. (Amsterdam, Leiden, The Hague, Utrecht, 1740) 4: 76–9.

[34] Hamou 2004, 203.

[35] It seems to me that you would have preserved your system from all these contradictions by simply claiming that the soul is a modification of matter, made in an unknowable way by the supreme being, who gave it the faculty to think, to understand etc, from which it follows that it is naturally mortal but made immortal by the divine will.

[36] Weil 1975, XIII–XIV.

[37] *Histoire des Ouvrages des Scavans*, Janvier 1706, 39.

concerning an extract from a refutation of Dodwell by the Rev. John Turner, saying:

> Il seroit bon que vous vissiez l'un & l'autre, pour en porter votre jugement. M. dodwel entendoit par le mot d'Esprit, suivant le langage du Nouveau Testament, un bienfait de dieu, distinct & séparé de l'ame, qui lui communique la vie, & l'immortalité. Cela mérite d'estre examiné.[38]

Cuenz followed his advice and finally tracked down a copy of the journal in the Genevan public library. This led to an exchange of views concerning Dodwell's opinion, in which Bouhier's prudence is once again in evidence, together with his willingness to accept in private apparently heterodox opinions which seem to him in accordance with his view of God. In reflections dating from the autumn of 1740 on Cuenz's latest letter, in which the latter criticized Dodwell's views in the belief that they were the same as Bouhier's, Bouhier notes:

> Je lui déclare, que je n'ai encore eu sur cela d'autre sentiment, sinon que dieu ne nous ayant point revelé, quelle est la nature de cet Etre; mais seulement qu'il est destiné à l'immortalité, je crois qu'on se tourmente inutilement, pour en savoir davantage, & pour pénetrer un mystére, que de grands génies ont jugé estre inaccessible à la raison humaine.
>
> Il est vrai, que j'ai ajouté, que s'il falloit prendre parti entre les différentes opinions, qui ont eté proposées sur cet article, il me semble que celle de M. dodwel seroit la plus aisée à défendre, n'imaginant aucun raisonnement, qui puisse la détruire solidement, & n'y trouvant rien de contraire aux vrais principes de notre Religion.[39]

Bouhier is retreating to a prudent position, accepting the limits of human reason and sticking to the truths taught by revelation. He agrees that matter as such cannot think, but continues to accept the hypothesis that God could give the capacity of thought to a particular type of organized subtle matter. The important point is to recognize that the soul is destined for immortality (fol. 316) and to avoid contradictions or anything that is opposed to Scripture or reason.

But he seems nevertheless to have gone even further in his acceptance of heterodox ideas. For Bouhier commented again on Cuenz's work (which incorporated several passages taken from their exchange) after its publication in 1742. He took particular issue with the remarks Cuenz added to a ten-page 'lettre de l'Anonyme' on Locke which he reproduced in his work. These pages were in fact a long extract from Voltaire's letter on Locke in his *Lettres philosophiques*.[40] Cuenz accused the author of following Lucretius rather than Locke, a judgement which Bouhier

---

[38] It would be a good idea to look at them both in order to judge for yourself. M. Dodwell understood by 'spirit', following the language of the New Testament, 'a blessing from God, distinct and separate from the soul, which gives it life and immortality'. That is worth examining.

[39] BNF mss. f. fr. 24410, fol. 314. I told him that I have not yet had any opinion on that subject other than that, as God has not revealed to us the nature of this being, but only that it is destined for immortality, I believe we torment ourselves in vain to know any more and to try to penetrate a mystery that great geniuses have judged inaccessible to human reason. § It is true that I added that if we had to decide between the different opinions that have been proposed on this issue, it seems to me that M. Dodwell's would be the easiest to defend, as I can imagine no reasoning that could solidly destroy it, and can find nothing in it contrary to the true principles of our religion.

[40] vol. I, 144–55.

considered far too harsh; according to him, such an accusation would only be valid if the anonymous author had denied that God could accord the soul immortality. He continued:

> Mais il ne dit rien de tel dans sa lettre. il ne raisonne, que sur la nature de notre Ame, indépendant de ce qu'il plait à Dieu d'ordonner de sa destinée. Ce dernier point appartient à la Révélation, sur laquelle il ne s'explique point, & par consequent il y auroit de l'injustice à suposer qu'il la nie. il parle en Philosophe & non en Théologien; & cependant c'est en cette derniere qualité, que vous le condamnez sans l'entendre.[41]

This distinction between the theologian and the philosopher was one that Bayle had made concerning Locke in his article 'Dicéarque'.[42] Nevertheless, the convergence expressed here between an antichristian thinker and a Christian apparently seeking sincerely to resolve to his own satisfaction questions of doctrine which posed problems and led to disagreements amongst philosophers, is interesting. It shows open-mindedness and a true willingness to examine Church doctrine in accordance with reason and accept opinions considered as irreligious (and even here as atheistic, as the reference to Lucretius shows) provided that they did not specifically contradict basic tenets of faith. While far from being an unbeliever, he could apparently agree with the claim of one of the most widely circulated of the clandestine irreligious works of the period (*L'Examen de la religion*) emphasizing the need to examine one's own religion.

This example also shows that one must be circumspect when judging heterodox texts or expressions of heterodox opinions in this period. They did not necessarily hide a desire to undermine religion, but could instead be an expression of the unease felt by some believers with certain doctrines which seemed unclear, contradictory or opposed to reason. These Christians followed with interest the lively and often invective-filled theological debates which could take place publicly among their Protestant neighbours and were reported extensively (sympathetically or unsympathetically) in learned journals. The example of Bouhier shows us that such Catholics were in addition willing to discuss such matters in private with Protestants in an open manner. At the same time, Bouhier was aware of the political dangers of allowing the open expression of such doubts and kept them private. He himself seems to have gone no further than such doubts and finally submitted to Church teachings and accepted the limitations of our reason. But the difficulty of expressing these doubts and examining one's religion openly could push some Catholics into the same camp as unbelievers, and undoubtedly helped to undermine belief and fuel the antichristian opinions that emerged more publicly in France from the middle of the century.

---

[41] BNF mss. f. fr. 24410, fol. 343: 'minute de la Réponse que j'ai faite le 20 Février 1742, à la lettre de Mr Cuenz du 3 du même mois'. But he says nothing of the sort in his letter. He is only reasoning on the nature of our soul independently of what fate God decides for it. This last point is part of revelation, on which he does not give an opinion, and thus it would be unjust to suppose that he denied it. He is speaking as a philosopher not a theologian, and yet it is in the latter capacity that you condemn him unheard.

[42] Pierre Bayle, article 'Dicéarque', rem. M, in *Dictionnaire historique et critique* (Paris: Desoer, 1820) 5: 512–5.

References

Benitez, Miguel. 1994. *La face cachée des Lumières. Recherches sur les manuscrits philosophiques clandestins de l'âge classique*. Oxford/Paris: Universitas.

Carayol, Elisabeth. 1984. *Thémiseul de Saint-Hyacinthe, 1684–1746*. Oxford: Voltaire Foundation.

Goldie, Mark. 1993. Priestcraft and the birth of Whiggism. In *Political discourse in early modern Britain*, ed. Nicholas Phillipson and Quentin Skinner, 209–31. Cambridge: Cambridge University Press.

Hamou, Philippe. 2004. L'opinion de Locke sur la "matière pensante". *Methodos* 4, translated in *John Locke: Critical assessments*, ed. Peter Anstey. London: Routledge 2006, 3: 176–212.

Israel, Jonathan I. 2000. *Radical enlightenment: philosophy and the making of modernity 1650–1750*. Oxford: Oxford University Press.

Jacob, Margaret C. 1981. *The radical enlightenment: pantheists, freemasons and republicans*. London: George Allen and Unwin.

Leighton, C.D.A. 2002. The religion of the Non-Jurors and the early British enlightenment: A study of Henry Dodwell. *History of European Ideas* 28: 247–62.

Rose, Craig. 1999. *England in the 1690s*. Oxford: Blackwell.

Thomson, Ann. 1981. Qu'est-ce qu'un manuscrit clandestin? In *Le matérialisme du XVIIIe siècle et la littérature clandestine*, ed. Olivier Bloch, 13–16. Paris: Vrin.

Thomson, Ann. 1988. L'homme-machine, mythe ou métaphore? *Dix-huitième Siècle* 20: 367–76.

Thomson, Ann. 2007. L'écho des débats théologiques anglais: Henry Dodwell et les journalistes. *La Lettre clandestine* 15: 77–92.

Thomson, Ann. 2008a. Un marginal de la République des Lettres: Caspar Cuenz. *Dix-huitième Siècle* 40: 29–42.

Thomson, Ann. 2008b. *Bodies of thought. Science, religion, and the soul in the early enlightenment*. Oxford: Oxford University Press.

Thomson, Ann. 2010. Animals, humans, machines and thinking matter, 1690–1707. *Early Science and Medicine* 15: 3–37.

Thomson, Ann. 2013. Toland, Dodwell, Swift and the circulation of irreligious ideas in France: what does the study of international networks tell us about the "Radical enlightenment"? In *An intellectual journey. Translation, circulation and adaptation in England, France and Ireland in the Enlightenment*, ed. Lise Andries and Frédéric Ogée. Oxford: Voltaire Foundation.

Trowell, Stephen. 1996. Unitarian and/or Anglican: The relationship of Unitarianism to the church from 1687 to 1698. *Bulletin of the John Rylands University Library of Manchester* 78: 77–101.

Vailati, Ezio. 1997. *Leibniz and Clarke: A study of their correspondence*. New York: Oxford University Press.

Wade, Iro O. 1975. Notes on the making of a *Philosophe*: Cuenz and Bouhier. In *Literature and history in the age of ideas*, ed. G.S.Williams Charles, 97–123. Columbus: Ohio State University Press.

Weil, Françoise. 1975. *Jean Bouhier et sa correspondance*. Paris: Université Paris-Sorbonne, CNRS, LA 96.

Yolton, John. 1991. *Locke and French materialism*. Oxford: Clarendon Press.

Index

A. Dunan-Page and C. Prunier (eds.), *Debating the Faith: Religion and Letter Writing in Great Britain, 1550–1800*, International Archives of the History of Ideas 209,
DOI 10.1007/978-94-007-5216-0,

MIX
Papier aus verantwortungsvollen Quellen
Paper from responsible sources
FSC® C105338

If you have any concerns about our products,
you can contact us on
ProductSafety@springernature.com

In case Publisher is established outside the EU,
the EU authorized representative is:
Springer Nature Customer Service Center GmbH
Europaplatz 3, 69115 Heidelberg, Germany

Printed by Libri Plureos GmbH
in Hamburg, Germany